EXERCISES IN CERTIFICATE MATHEMATICS 2

G M Hocking B.Sc.
Head of Mathematics, Kingsfield School, Bristol
Assistant Examiner (O Level), London Board
Chief Examiner (CSE), South Western Examination Board

MACMILLAN

First published 1974

Published by
MACMILLAN EDUCATION LIMITED
London and Basingstoke

Associated companies and representatives throughout the world

Printed in Great Britain by
A. WHEATON AND CO.
Exeter

Preface

This book – and its companion 'Exercises in Certificate Mathematics 1' – is for C.S.E. and O Level students. It is planned for use in the final two years of their certificate course and intended to enrich and extend their mathematical experience and to develop confidence in the application of mathematical techniques and concepts through challenge and success.

The questions in the two books cover almost all the requirements of certificate examinations in mathematics at C.S.E. and O Level. They have been used in pilot schemes with candidates whose abilities vary from somewhat below to substantially above average and their content and form modified in the light of this experience.

Book 2 supplements and complements the work of Book 1, reviewing the essential topics and techniques of Sets, Calculating – including the use of tables, slide rules and hand calculators – the Algebra of Real Numbers, Relations, Graphs and the Geometry and Trigonometry of two and three dimensional figures. This review of basic mathematics is augmented by extensive exercises on the now well-established topics of Statistics, Matrices and Vectors.

The contents are designed to be both instructive and imaginative. The drudgery of over-repetitive questions has been avoided and the emphasis placed on providing a variety of question types and topics, that simultaneously stimulate and educate, in a lively and enlightening style.

It is not intended that any one student should attempt every question or indeed every exercise. For revision purposes, it is sufficient to try every other question in a relevant exercise, or every third question, or even every prime numbered question, filling in with further examples where the student is in difficulty or on unfamiliar ground. Such a procedure will provide a sufficient cross-section of experience in a particular topic. Students should be encouraged to make up questions for themselves, using the given examples as prototypes.

As ever, the instruction and guidance of a stimulating and understanding teacher is vital. For such a teacher the book provides, in compact form, a wealth of questions which he may select and readily augment to meet the needs of his particular pupils.

My thanks are due to staff and students who with zest and care have checked answers and suggested improvements in wording. Finally, I must record my gratitude to Mr Tim Pridgeon for his quiet faith and gentle guidance throughout its preparation.

G M HOCKING

Contents

Sets

Exercise 1

1 Say whether the following statements about the sets in Fig. 1.1 are true or false.
(a) $Q \subset P$, (b) $P \cup R = R$, (c) $P \cap Q = Q$,
(d) $Q \cap P' = \phi$, (e) $P' \cap Q' = \phi$, (f) $R \subset (Q \cup P)$,
(g) $(R \cup Q') \subset P$.

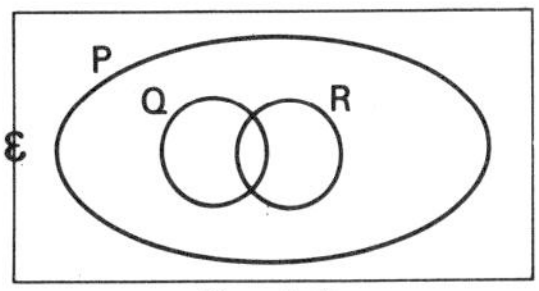

Fig. 1.1

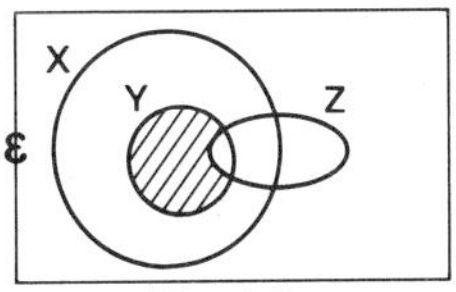

Fig. 1.2

2 Make four copies of the Venn diagram in Fig. 1.2 and shade (a) $(X \cup Z)$, (b) $X \cup Y \cup Z'$, (c) $X \cap Y' \cap Z$, (d) $X \cap (Y \cup Z)$.
Express the shaded region in terms of X,Y and Z.

3 $\mathscr{E}$ = {whole numbers from 1 to 20 inclusive}
N = {whole numbers between 2 and 18}
M = {multiples of 4}
P = {7, 11, 13}.
Say whether the following statements are true or false.
(a) $P \cap M = \phi$, (b) $P \cup N = P$, (c) $P \cap N = P$,
(d) $M \subset N$, (e) $N' = \{1, 2, 18, 19, 20\}$, (f) $n(M) = 5$,
(g) $\{3, 6, 9\} \subset M'$, (h) $n(P') = 17$.

4 Using the numbers in Fig. 1.3 find $n(P)$, $n(R)$, $n(P \cap Q)$, $n(Q \cup R)$, $n(P \cap Q')$, $n(\mathscr{E})$, $n[(P \cup Q) \cap R]$, $n(P \cap R')$, $n[P \cup (Q \cap R)]$.

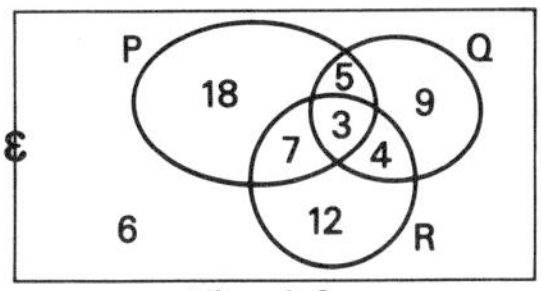

Fig. 1.3

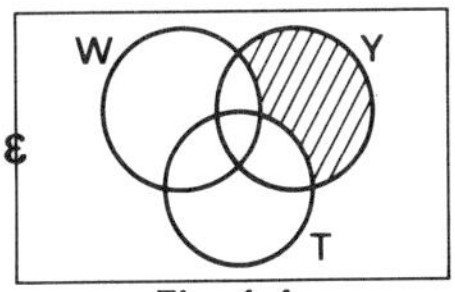

Fig. 1.4

5 Make four copies of the Venn diagram in Fig. 1.4 and shade (a) $W \cap Y \cap T$, (b) $(W \cup Y) \cap T$, (c) $W \cup (Y \cap T)$, (d) $(W \cup Y)' \cap T$.
Express the shaded region in terms of W, Y and T. If W = {plants that flower in Winter}, T = {plants taller than 1 m} and Y = {plants with yellow flowers}, what is the meaning of $(W \cup Y) \cap T$?

6 $\mathscr{E}$ = {playing cards}, C = {court cards} = {King, Queen, Jack}, H = {honour cards} = {Ace, King, Queen, Jack, Ten}.

(a) Illustrate these sets by means of a Venn diagram. (b) List the elements in $C \cap H$, $C' \cap H'$, $C' \cap H$.

7 Draw Venn diagrams to illustrate the following arguments. In each case say whether the conclusion is logically valid or invalid.

(a) All car owners pay taxes. Fred owns a car. Therefore Fred pays taxes.

(b) Tubby people are jolly. Donkey owners are jolly. Therefore donkey owners are tubby.

(c) Lavender is fragrant. All perfumes are fragrant. Therefore lavender is a perfume.

(d) Persistent T.V. viewers are lazy. Jessica is lazy. She watches too much T.V.

(e) Professional footballers are well paid. Alf Thyme is a professional footballer. Alf is well paid.

(f) Jugglers are nimble. Frederico is nimble. Therefore Frederico is a juggler.

(g) No animal in this zoo goes hungry. This cat is starving. Therefore this cat does not belong to this zoo.

(h) All chocolate biscuits are sweet. Some sweet biscuits are wrapped in silver paper. Therefore some chocolate biscuits are wrapped in silver paper.

8 Use the relation $n(A \cup B) = n(A) + n(B) - n(A \cap B)$ to test whether the following statistics are consistent.

	$n(A \cup B)$	$n(A)$	$n(B)$	$n(A \cap B)$
(a)	50	35	25	10
(b)	144	86	42	16
(c)	72	38	48	8
(d)	95	58	62	25
(e)	150	56	65	29
(f)	84	54	44	14

Correct those that are inconsistent – assuming that the error is in the third column only.

9 It was reported that out of 35 sixteen-year-olds, 22 had Saturday jobs, 18 had a summer holiday job, 12 had both and 4 had neither.

Explain why this is impossible. Correct the figures if the error is in (a) the Saturday jobs, (b) the holiday jobs.

10 Yasha the spy reported 36 pilots stationed at aerodrome X7. 20 pilots, he said, could fly bomber B1, 18 could fly bomber UB2. There were 8 pilots, Yasha reported, able to fly B1 and UB2. Is there any reason to suspect this information?

11 On St. Valentine's Day, 26 girls received at least one card, 10 received at least two cards and 4 received three or more cards. How many received one card only?

12 Of 20 old cars tested, 12 had faulty brakes, 15 had faulty lights and 8 had faulty steering. 3 cars had all three of these faults. None were fault-free. How many cars had exactly two faults?

13 72 athletes took part in a trial. There was a sprint (S), a long jump (J) and a distance race (D).

No one took part in both S and D. 16 took part in J only. 5 took part in J and D and twice as many as this entered both S and J. In all there were 8 more sprinters than distance runners.

Find the number of sprinters.

14 The Clearwater Aquatic Club has 70 members all of whom take part in at least one of the Club's activities of water skiing (K), sailing (S) and swimming (M).

There are 18 members of the club whose only interest is skiing. 25 take part in just two activities of whom 6 ski and sail, 9 ski and swim. 4 members are so full of energy that they join in all three activities.

Find the total number who sail, if this is 4 less than the total number who swim. How many merely swim?

15 12 members of a class had football boots, 10 had hockey boots and 15 had rugby boots.

8 had football and rugby boots and 1 of these had hockey boots as well. 3 had both hockey and rugby boots but no football boots. 1 had football boots only.

Find how many had (a) football and hockey boots, (b) hockey boots only, (c) rugby boots only.

How many were there in the class, if 4 had no boots at all?

16 Figure 1.5 shows an alternative to a Venn diagram for illustrating the sets P, Q and R. Copy the figure and find $n(P)$, $n(R)$, $n(P \cap Q)$, $n(P \cup Q \cup R)$, $n(P \cap Q \cap R)$.

Check that $n(P \cup Q \cup R) = n(P) + n(Q) + n(R) - n(P \cap Q) - n(Q \cap R) - n(R \cap P) + n(P \cap Q \cap R)$.

17 Use the information in Fig. 1.5 to find $n(P \cap Q')$, $n(R')$, $n(P' \cup Q')$, $n(P \cap Q')$.

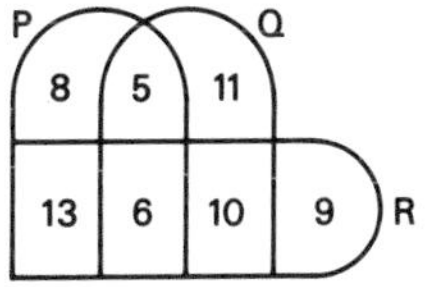

Fig. 1.5

18 The 72 employees of the Nutcracker Engineering Co. use one or more of four types of machine W, X, Y and Z.

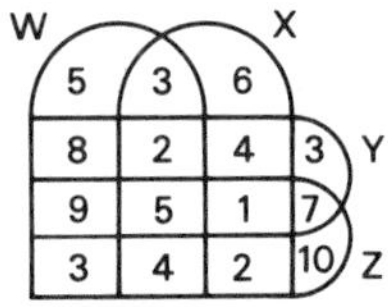

Fig. 1.6

Use Fig. 1.6 to find the number who operate (a) machine W only, (b) machine X at least, (c) one machine only, (d) all four machines, (e) machines X and Y at least, (f) just three machines.

19 Eight punched cards are used to represent eight sets. The cards are numbered 0 to 7 in binary form with a hole for 0 and a slot for 1.

(a) How would set number 5 be represented? (b) How would sets number 4, 7, 0 be represented? (c) How many spaces are needed for holes/slots?

20 The cards in question 19 are selected by sliding needles through the holes or slots. Explain how to find (a) the odd numbered sets, (b) set number 5, (c) set number 7.

21 The eight cards in question 19 are shuffled. Explain how to arrange the cards in numerical order using the needles.

22 How many different sets can be represented on cards with spaces for (a) 3 holes, (b) 4 holes, (c) 6, (d) 8, (e) 10, (f) 20 holes?

23 The following information is to be recorded on punched cards about members of a form. (1) Boy or girl.
(2) House – out of four houses. (3) Taking lunch at school or not. (4) Living within 1 mile of school or more than this.

(a) What is the least number of spaces needed for items 1 and 3? (b) Explain how to economise in spaces for item 2.
(c) What is the least number of spaces needed altogether?

24 Thirty cards are prepared, as in question 23, for the members of a class. Explain how to select using a needle (a) the boys – assume 0 for a boy, (b) the girls in house 01, (c) the pupils in house 11 who stay for lunch – assume 0 for staying, (d) the pupils who live more than a mile away and still go home for lunch – assume 0 for within 1 mile.

25 Peter, Quintin, Ray, Silent Sam and Tiddley Tim are the five members of a local pop group. P is free on the evenings of

Monday, Wednesday and Saturday. Q works Tuesday and Thursday evenings. R can play on any day except Sunday and Monday. S is free Monday, Wednesday, Saturday and Sunday. T is unemployed and is available throughout the week.

Show punched cards for each member using a space for each day of the week, 0 for not available and 1 for available.

(a) On which nights of the week can the whole group play? (b) Which is the worst night for rehearsals out of Monday, Tuesday and Wednesday? (c) Which member is least available? (d) Could the group play more frequently if they became a quartette? Who should drop out? (e) If T is too unreliable to take telephone bookings, who is available to record bookings on (1) Tuesdays, (2) Fridays?

26 There are four interpreters on the staff of the Stilton Hotel. Alex speaks English, French and Italian. Boris speaks Russian, English and German. Carmen speaks Spanish, French and English. Dimitri speaks all the above languages except Russian and Italian.

Sketch punched cards to show the languages spoken by the four using a space for each language and 1 for 'speaks'.

(a) Which language is spoken by all four? (b) Who can translate French into English? (c) How many languages are spoken by 2 or more interpreters? (d) One hotel guest spoke Italian only and another Russian only. Which interpreters are needed to enable the Italian to negotiate a business deal with the Russian? (e) Which interpreter is inessential?

27 Sketch Venn diagrams to illustrate
(a) $(A \cap B) \cap C = A \cap (B \cap C)$,
(b) $(A \cup B) \cup C = A \cup (B \cup C)$,
(c) $(A \cup B) \cap C \neq A \cup (B \cap C)$,
(d) $(A \cap B)' = A' \cup B'$,
(e) $A \cap (B \cup C) = (A \cap B) \cup (A \cap C)$.

28 Which of the following statements about sets are always true? (a) $P \cap Q = Q \cap P$, (b) $P \cup Q \cup R = P \cup R \cup Q$, (c) $P \cup Q' = Q \cap P'$, (d) $(P \cup Q)' = P' \cap Q'$, (e) $P \cup P = P$, (f) $Q \cap Q = \phi$, (g) $P \cup (Q \cap R) = (P \cup Q) \cap R$, (h) $P \cap (Q \cup R) = (P \cap Q) \cup (P \cap R)$.

29 Draw Venn diagrams and hence simplify the following expressions.
(a) $(P \cup P')$, (b) $(Q \cap Q')$, (c) $(P \cap Q') \cup Q$, (d) $(P \cap Q) \cup (P \cap Q')$, (e) $(P \cap \mathscr{E})$, (f) $(Q \cup \mathscr{E})$.

30 Simplify: (a) $P \cap \phi$, (b) $Q \cup \mathscr{E}$, (c) $R \cup (R' \cap S)$, (d) $P \cap (Q' \cup P)'$, (e) $\{(P \cap \phi) \cup P'\}'$.

31 $A \cup B'$ and $A' \cap B$ are an example of a pair of Dual Expressions, in which A is replaced by A′, $\mathscr{E}$ by ϕ and $\cap$ by $\cup$.

Illustrate these two expressions by Venn diagrams and explain the dual relation between them.

Write the dual of the following expressions and illustrate the dual pairs by Venn diagrams.

(a) $P \cup Q$, (b) $P' \cup Q'$, (c) $P' \cap Q$.

32 Write the dual of the expression $P \cup Q \cup R'$ and illustrate the dual pair by means of a Venn diagram.

33 Repeat question 32 for (a) $(P \cup Q') \cap R$, (b) $P \cup (Q \cap R)$.

34 $P - Q$ (read as P difference Q) is defined as the set of elements in P that are not in Q.

$P = \{2\ 4\ 6\ 8\}$, $Q = \{4\ 9\ 16\}$, $R = \{3\ 6\ 9\}$.

Find $P - Q$, $Q - P$, $Q - R$, $R - Q$, $P - (Q - R)$, $(P - Q) - R$.

35 (a) Draw a Venn diagram to illustrate $R - S$. (b) Draw another Venn diagram to illustrate $R \cap S'$. Comment on the relation between the two diagrams. (c) Illustrate $S - R$ by a Venn diagram. (d) Find an alternative expression for $S - R$. (e) Illustrate $R - (S - T)$ and $(R - S) - T$. What can you say about the expression $R - S - T$?

36 $P \mathbin{\Delta} Q$ is called 'the Symmetric Difference of P and Q'. It consists of those elements of P that are not in Q together with those elements of Q that are not in P.

Use the sets P, Q and R from question 34 and write down $P \mathbin{\Delta} Q$, $Q \mathbin{\Delta} P$, $(P \mathbin{\Delta} Q) \mathbin{\Delta} R$ and $P \mathbin{\Delta} (Q \mathbin{\Delta} R)$.

37 Draw Venn diagrams to show that

(a) $P \mathbin{\Delta} Q = (P \cup Q) - (P \cap Q)$,

(b) $P \mathbin{\Delta} Q = (P \cup Q') \cup (Q \cap P')$.

Calculating

Exercise 2

Estimate, then use tables or slide rules to answer the following questions.

1 Evaluate:

(a) $19{\cdot}6 \times 17{\cdot}2$ (b) $1{\cdot}96 \times 17{\cdot}2$ (c) $58{\cdot}4 \times 32{\cdot}2$

(d) $58{\cdot}4 \div 32{\cdot}2$ (e) $23{\cdot}2 \div 58{\cdot}4$ (f) $42{\cdot}4 \div 16{\cdot}8$
(g) $16{\cdot}8 \div 42{\cdot}4$ (h) $16{\cdot}8 \div 0{\cdot}424$ (i) $\pi \times 15{\cdot}7^2$
(j) $\pi \times 18{\cdot}2^2 \times 41{\cdot}6$ (k) $4\pi \times 17{\cdot}3^2$ (l) $\frac{4}{3}\pi \times 8{\cdot}3^3$
(m) $\sqrt{18{\cdot}4}$ (n) $\sqrt{184}$ (o) $\sqrt{0{\cdot}184}$ (p) $\sqrt{19{\cdot}2^2 + 15{\cdot}9^2}$
(q) $\sqrt{19{\cdot}2^2 - 15{\cdot}9^2}$ (r) $\sqrt{19{\cdot}2 \div 15{\cdot}7}$ (s) $\sqrt[3]{56{\cdot}8}$
(t) $\sqrt[3]{0{\cdot}568}$ (u) $\frac{1}{1{\cdot}84}$ (v) $\frac{1}{18{\cdot}4}$ (w) $\frac{1}{0{\cdot}84}$ (x) $\frac{1}{1{\cdot}66}^2$
(y) $\frac{1}{0{\cdot}166}^2$ (z) $\frac{1}{0{\cdot}66}^2$

2 Calculate the areas shown in Fig. 2.1, with lengths in cm.

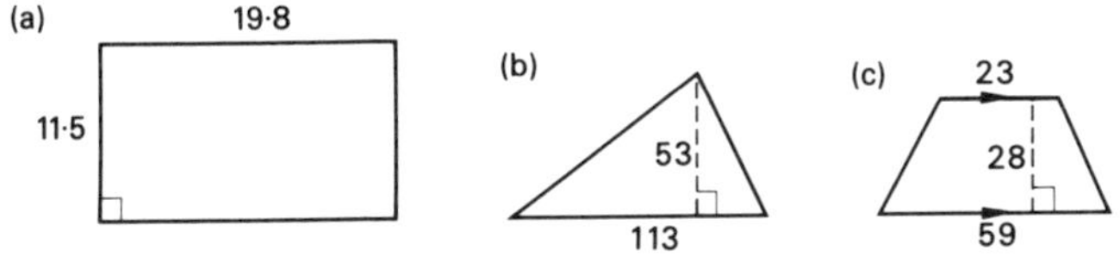

Fig. 2.1

3 Use 1 m = $1{\cdot}09$ yd to complete the table below.

yd					120	1 760	440	2 000
m	100	48	225	1 500				

4 Take 1 kg = $2{\cdot}02$ lb to find the corresponding masses.

kg	28	44	58	700				
lb					112	345	500	2 240.

5 Find the value of: (a) $18{\cdot}2 \times 13{\cdot}6 \times 1{\cdot}97$,
(b) $18{\cdot}2 \times 13{\cdot}6 \div 1{\cdot}97$, (c) $0{\cdot}182 \times 13{\cdot}6 \div 0{\cdot}0197$,
(d) $\left(\frac{15{\cdot}8}{1{\cdot}95 \times 23{\cdot}6}\right)^2$ (e) $\sqrt{\frac{1{\cdot}95 \times 23{\cdot}6}{15{\cdot}8}}$ (f) $\left(\frac{0{\cdot}846}{0{\cdot}062}\right)^3$.

6 (a) Find the value of $5{\cdot}6^2$, $15{\cdot}6^2$, 156^2, $0{\cdot}56^2$.
(b) Find the areas of squares with sides 29, 53, 106, 10·6, 20·6 and 187 cm.
(c) Use $A = \pi r^2$, to find the areas of circles with radius 5·4, 6·3, 8·2, 82, 17·6 and 176 cm.

7 (a) Evaluate 12^3, 23^3, $2{\cdot}3^3$, $4{\cdot}5^3$, 45^3, 131^3.
(b) Find the volumes of cubes with sides 13, 17, 28, 37, 4·8, 8·6 and 12·4 cm.
(c) Use $V = (4/3)\pi r^3$, to find the volumes of spheres with radius 3·8, 4·6, 7·3, 12·9 and 25·6 cm.

8. (a) Find $\sqrt{177}$, $\sqrt{1\,770}$, $\sqrt{232}$, $\sqrt{23{\cdot}2}$ $\sqrt{69{\cdot}3}$ $\sqrt{6{\cdot}93}$.
(b) Find the sides of the squares with areas 18, 42, 75, 750, 1 200, 3 670 cm².

9 (a) Evaluate $\sqrt[3]{57}$, $\sqrt[3]{57{\cdot}7}$, $\sqrt[3]{135}$, $\sqrt[3]{1\,350}$, $\sqrt[3]{17{\cdot}8}$, $\sqrt[3]{1\,780}$.
(b) Find the side of cubes with volumes 166, 1 606, 16·06 and 160·6 cm³.

10 Use the formula $C = \pi d$, to find the circumferences of tyres with diameter 31, 48, 56, 156 and 1·56 cm.

11 Use $A = 2\pi r\ (r + h)$ to find the total surface area of the following cylinders.

Radius r cm	5·4	5·8	11·6	19·8
Height h cm	16	20	22·2	31·5

12 Convert the following marks out of 55 to percentages.
Mark: 23, 51, 42, 33, 18, 52, 28, 49.

13 Express as percentages the following marks out of 72.
Mark: 44, 55, 63, 15, 37, 69, 59, 48.

14 Calculate the average speed in km/h of motor cycles that lap a 2·38 km circuit in 82·2, 79·5, 78·7, 77·7 and 76·3 seconds.

Use a hand calculator for the remaining questions in this exercise.

15 Compute (a) 12 345 + 54 321, (b) 54 321 − 12 345, (c) 3 113 + 1 331, (d) 3 113 − 1 331, (e) 3 113 × 1 331, (f) 3 113 ÷ 1 331, (g) 121^2, (h) $9\,898^2$, (i) 40 000 ÷ 99.

16 R = 18 025 and S = 7 602. Find R + S, R − S, 3R + 2S and 5R − 3S.

17 P = 115·32 and Q = 84·89. Find P + Q, P − Q, 2P + 3Q and 3P − 2Q.

18 L = 119 and M = 155. Find LM, L^2, M^2, L^2M, L ÷ M and M ÷ L.

19 S = 58·92 and T = 34·4. Find ST, S^2, ST^2, $(ST)^2$, S ÷ T and T ÷ S.

20 x = 587, y = 316 and z = 119.
Compute x + y + z, x − y + z, x − y − z, xy, yz, xyz, x^2, $y^2 + z^2$, x ÷ y, y ÷ z, z ÷ x.

21 p = 53·9, q = 24·6 and r = 2·45.
Evaluate p + q + r, pq, qr, pqr, p^2, pq^2, $(pq)^2$, p ÷ q, q ÷ r, r ÷ p.

22 Calculate the length of fencing required to surround each of the fields with sides (a) 189, 346, 213 and 304 m, (b) 546, 412, 739 and 648 m, (c) 1 054, 1 121, 987 and 834 m, (d) 1 213, 969, 882, 1 435 and 765 m.

23 Find the total cost of each of the following bills.

(a)	(b)	(c)	(d)
84 @ £0.82	56 @ £0.76	72 @ £0.34	44 @ £1.36
272 @ £0.16	48 @ £0.33	36 @ £0.47	55 @ £2.11
140 @ £1.25	32 @ £1.17	145 @ £0.58	32 @ £1.17
135 @ £3.77	88 @ £2.22	54 @ £1.31	84 @ £2.46

24 Compute the cash price of freezers listed at £88, £96, £110 and £125 when there is a cash discount of 15%.

25 Repeat question 24 for discounts of (a) $17\frac{1}{2}$%, (b) $22\frac{1}{2}$%.

26 Find, to the nearest penny, 6% of £45, 8% of £22.40, 12% of £5.55, 16% of £58.85 and 35% of £1 006.

27 The speed 1 knot is approximately 1·8532 km/h. Convert the following airspeeds to km/h, giving answers to the nearest whole number. Airspeeds 175, 358, 635, 1 354 and 1 831 knots.

28 The exchange rate for dollars stood at $2.47 to £1. Convert to dollars, £97, £183, £238, £716, £1 525 and £108 601 – answers to the nearest cent.

29 Repeat question 28 for Pesetas at 13·16 Pesetas to the pound. Give answers to the nearest Peseta.

30 Explain how to economise in computing (a) $169 \times 14{\cdot}7$, (b) $15{\cdot}97^2$, (c) $12{\cdot}3^4$.

31 A racing circuit is 3·37 km in length. Set up a programme to compute the average speed of cars, in km/h, lapping the circuit in *s* seconds. Use your programme to find the average speeds corresponding to lap times of 101·8, 98·7, 93·6 and 85·3 seconds.

32 Devise a programme to compute y, where $y = x^2 - 3x + 2$. Find y when x equals 17, 1·5, 15·9.

33 Repeat question 32 for (a) $y = x^2 + 4x - 6$ and $x = 24$, 9·5 and 18·6, (b) $y = 2x^2 - x + 3$ and $x = 8$, 28 and 2·8.

34

Game	1	2	3	4
Attendance	26 076	18 320	23 545	21 482

The table above shows the attendance at four home games of a football club. Calculate: (a) the total attendance, (b) by how much the maximum attendance exceeded the minimum attendance, (c) the average attendance, (d) the takings, in £s, for each game if the admission fees averaged 34p per person.

Suggest a way in which you could check your answer to part (d).

35 A jet liner makes a one way Atlantic flight of 4 850 km each day for six days of every week. Its average times are (1) west to east 5 h 15 min, (2) east to west 5 h 45 min. Compute: (a) the total distance flown per week, (b) the total distance flown per year, (c) by how much this distance exceeds $1\frac{1}{2}$ million km, (d) the average speed (i) east to west, (ii) west to east, (e) the total time spent in the air each year.

36 The data below gives details of five circus elephants.

Weight of elephant (kg)	1 325	934	914	873	576
Daily food ration (kg)	57	47	51	46	36

Compute: (a) the difference in weight between the heaviest and the lightest elephants, (b) the total weight of the five, (c) the total weight of food eaten per week, (d) the average food consumption of an elephant per day.

Find which elephant has the greatest $\dfrac{\text{Weight of Food}}{\text{Weight of Elephant}}$ ratio.

37 Some old farm documents gave the areas of fields in acres as:

Field	Great Meadow	Lower rise	Upper rise	Sandy bottom	Hill top
Area (acres)	8·4	3·2	6·5	5·7	4·8

(a) Explain how to convert these areas to hectares if 1 acre = 0·405 hectares. (b) Convert the five areas to hectares to two decimal places. (c) Find the total area in hectares. (d) Suggest a method of checking your answer and carry out the check. (e) How many acres are there in one hectare – to two decimal places?

38 The table below shows the attendances at the Mudbank Swimming Baths during one week.

Day	Mon	Tues	Wed	Thur	Fri	Sat	Sun	Admission
Adults	49	125	148	137	249	423	306	$12\frac{1}{2}$p
Children	206	254	304	283	321	385	243	6p

Compute (a) the total daily attendances, (b) the total attendance for the week. (c) Check your answers and explain your method of checking. (d) Find the total takings for the week.

39

Year	1963	1968	1969
No of books published	26 023	31 420	32 393

(a) Find the total number of books published in the three

years. (b) How many more books were published in 1969 than than in 1963? (c) What percentage increase is this?

40 Compute, to the nearest penny, the Compound Interest on (a) £840 for 4 years at 6% (b) £560 for 4 years at $4\frac{1}{2}$% (c) £780 for 4 years at $7\frac{1}{2}$% (d) £1 020 for 5 years at 3%.

41 The value of a new house is estimated to increase by 6% each year. Find the value after 3 years of houses that originally cost £4 800, £6 500 and £8 400. Give answers to the nearest £25.

42 A caravan is bought for £570 and depreciates by 15% each year. Compute its value after 4 years, to the nearest £10.

43 Repeat question 42 for (a) a £1 300 car depreciating by 20% for each of 4 years, (b) a £15 000 yacht depreciating by $17\frac{1}{2}$% for each of 4 years.

44 An alligator increases its length by 60% every 6 months. Starting with 24 cm of alligator at the beginning of 1972 how long is it at the end of 1974?

Indices and Logarithms

Exercise 3

1 Write out in full and find the value of: 2^4, 3^3, 5^6, 6^5 10^7 and 12^3.

2 Which has the largest value: 2^5, 5^2 or 3^3?

3 (a) Arrange in ascending order 3^4, 4^3, 5^3 and 3^5.
(b) Arrange in descending order 6^2, 2^6, 3^4 and 4^5.

4 When multiplied out completely what is the units figure for 10^8, 5^{13}, 9^{15}, 3^{10}, 4^6, 7^7 and 6^n, where n is any natural number?

5 Evaluate: $2^8 \div 2^5$, $3^{19} \div 3^{15}$, $4^7 \div 4^5$, $5^{12} \div 5^{10}$, $6^7 \div 6^6$, $9^5 \div 3^6$, $15^4 \div 5^4$, $18^{10} \div 9^{10}$.

6 Calculate, in the shortest way possible: $43^4 \div 43^3$, $18^3 \times 18^2 \div 18^6$, $\sqrt{2^6 \cdot 8^4}$, $(56^2)^3 \div 28^7$.

7 Simplify: $a^3 \times a^2$, $b^5 \div b^4$, $c^2 \times c^3 \times c^4$, $2(d^2)^3$, $(2e^3)^2$, $f^3 \times f^4 \div f^7$, $\sqrt{g^6}$, $\sqrt[3]{h^9}$.

8 Simplify: $ab^2 \times a^2b$, $b^4c^3 \div b^2c$, $(cd^2)^2$, $(de)^2 \times d^3e^4 \div d^5e^6$, $a^2bc^2 \div ab^2c$, $(a^2b \div b^2c) \times ac^2$, $a^2b \div (b^2c \times ac^2)$.

9 If $n = 2^6 \times 5^{12}$, find in index form: $\sqrt{n}$, n^2, n^3 $\sqrt[3]{n}$.

10 If $p = 2^4 \times 7^3$, find the smallest whole number that p must be multiplied to make it (a) a perfect square (b) a perfect cube (c) a multiple of 42.

11 Evaluate (a) $36^{\frac{1}{2}}$, $36^{-\frac{1}{2}}$, $36^{1\frac{1}{2}}$ (b) $125^{\frac{1}{3}}$, $125^{-\frac{1}{3}}$, $125^{\frac{2}{3}}$ (c) $64^{\frac{1}{3}}$, $64^{\frac{1}{6}}$, $64^{-\frac{1}{3}}$, $64^{-\frac{2}{3}}$ (d) $81^{\frac{1}{2}}$, $81^{\frac{1}{4}}$, $81^{\frac{3}{4}}$, $81^{-\frac{1}{2}}$, $81^{1\frac{1}{2}}$ (e) $(\frac{49}{64})^{\frac{1}{2}}$, $(\frac{49}{64})^{-\frac{1}{2}}$, $(\frac{27}{64})^{\frac{1}{3}}$, $(\frac{27}{64})^{-\frac{2}{3}}$ (f) $(0{\cdot}01)^{\frac{1}{2}}$, $(0{\cdot}04)^{\frac{1}{2}}$, $(0{\cdot}04)^{-\frac{1}{2}}$, $(0{\cdot}008)^{\frac{1}{3}}$, $(0{\cdot}008)^{-\frac{2}{3}}$.

12 Simplify: $49^{\frac{1}{2}}$, $100^{\frac{1}{2}}$, $81^{-\frac{1}{4}}$, $32^{\frac{2}{5}}$, $49^{1\frac{1}{2}}$, $4^{-\frac{1}{2}}$, $36^{-1\frac{1}{2}}$, $64^{-\frac{1}{6}}$, $64^{\frac{5}{6}}$, $0{\cdot}001^{\frac{1}{3}}$, $0{\cdot}064^{-\frac{1}{3}}$.

13 Find x if: $2^x = 32$, $4^x = 32$, $8^x = 32$, $3^x = 27$, $3^x = 1/27$, $9^x = 27$, $10^x = 1\,000$, $10^x = 0{\cdot}001$, $10^x = 0{\cdot}01$.

14 Find x if: $2^x = 16$; $2^{x+2} = 16$; $3^x = 81$; $3^{2x-1} = 9$; $3^{3x} = 9^{x+2}$; $2^{4x} = 8^{x+1}$; $4^{x-1} = 8^{6-x}$.

15 Find a and b if $2^{a+b} = 32$ and $3^{a-b} = 27$.

16 Solve: $x - 4x^{\frac{1}{2}} + 3 = 0$; $x + x^{\frac{1}{2}} - 6 = 0$; $x^{\frac{2}{3}} - 3x^{\frac{1}{3}} + 2 = 0$.

Unless otherwise stated, the logarithms in the following questions are to the base ten.

17 If $\log x = 1{\cdot}246$, find: $\log x^2$, $\log\sqrt{x}$, $\log \frac{1}{x}$, $\log 10x$, $\log 100x$.

18 Repeat question 17 for (a) $\log x = 2{\cdot}48$, (b) $\log x = 0{\cdot}362$ (c) $\log x = \bar{1}{\cdot}64$.

19 If $p = \log x$, express in terms of p: $\log x^3$, $\log\sqrt{x}$, $\log \frac{1}{x}$, $\log 10x$, $\log \frac{x^2}{100}$.

20 If $s = \log 3$ and $t = \log 2$, express in terms of s and t, $\log 9$, $\log \frac{1}{2}$, $\log 6$, $\log 12$, $\log 27$, $\log 48$, $\log 1{\cdot}5$, $\log 4{\cdot}5$.

21 If $m = \log 4$ and $n = \log 5$, express in terms of m and n, $\log 16$, $\log 2$, $\log 25$, $\log 1/5$, $\log 20$, $\log 80$, $\log 1/50$.

22 Find: $\log_3 27$, $\log_3 81$, $\log_{81} 3$, $\log_9 27$, $\log_4 16$,

$\log_4 64$, $\log_4 1/64$, $\log_4 2$, $\log_4 8$, $\log_5 25$, $\log_5 125$, $\log_5 1/125$, $\log_5 0{\cdot}04$.

23 Express in terms of $\log x$: $\log x^3 + \log x^2$; $\log \sqrt{x} + \frac{1}{2} \log x^3$; $\log x^5 - \log x^2$; $\log x^4 + \log x^5$.

24 Simplify: $\log 200 + \log 5$; $3 \log 2 + 2 \log 5 - \log 2$; $\log 16 - 4 \log 2$; $3 \log 5 + \log 8$; $6 \log 2 - 2 \log 8$.

25 Find x if: $\log_5 x = 2$; $\log_3 x = 4$; $\log_2 x = 5$; $\log_4 x = 2{\cdot}5$; $\log_9 x = \frac{1}{2}$; $\log_8 x = \frac{1}{3}$; $\log_8 x = -\frac{1}{3}$.

26 What is the value of x if: (a) $\log x = 2 - \log 5$; (b) $\log 2x = 1 + \log 4$; (c) $\log x = 1 - \log 2$; (d) $2 \log x = \log (3x - 2)$; (e) $\log \frac{4}{x} = 2$?

The Algebra of Real Numbers

Exercise 4

1 Expand the following: (a) $4(3x + 2)$, (b) $3(2x - 5)$, (c) $4x(x - 3)$, (d) $3x(1 - 2x)$, (e) $5(x + 3) - 2(x + 4)$, (f) $5x(x + 2) - 3x(x - 4)$, (g) $(2x + 5)(3x - 2)$, (h) $(4x - 1)(x + 3)$, (i) $(x - 1)^2$, (j) $(3x - 2)^2$, (k) $(2x + 5)(2x - 5)$, (l) $(x + 1)^2 - (x - 2)^2$.

2 Factorise: (a) $p^2 + p$; $3q^2 + 15q$; $4r^2 - 36r^3$; $\pi R^2 h + \pi r^2 h$; $\frac{1}{2}sh + \frac{1}{2}th$ (b) $p^2 + pq + p + q$; $r^2 - rs + 3r - 3s$; $ce + 2cf + de + 2df$; $ac + bc + 2bd + 2ad$ (c) $x^2 + 9x + 20$; $x^2 - 9x + 18$; $8 + 2x - x^2$; $15 + 2x - x^2$; $2x^2 + 7x + 3$; $3x^2 + 10x - 8$ (d) $\mathrm{r}^2 - 36$; $s^2 - 144$; $9t^2 - 49$; $3p^2 - 75$; $128 - 2q^2$.

3 Factorise and evaluate: (a) $18{\cdot}3 \times 16 + 18{\cdot}3 \times 4$, (b) $52{\cdot}7 \times 18{\cdot}5 + 1{\cdot}5 \times 52{\cdot}7$, (c) $14^2 - 13^2$, (d) $418^2 - 318^2$, (e) $59 \times 21 + 21^2$, (f) $\pi \times 25 + \pi \times 17$, (g) $\pi \times 95 - \pi \times 88$. (Take $\pi = 22/7$)

4 Simplify: (1) $a^2b \times ab^2$; $2(b^2)^3$; $(2b^2)^3$; $\sqrt{9c^4}$; $10cd^2 \div 5c^2d$; $(d^2e \div de^2)^2$; $\sqrt{fg^3} \times gf$ (2) $\frac{1}{3a} + \frac{1}{4a}$; $\frac{1}{3b} - \frac{1}{4b}$; $\frac{1}{3c^2} + \frac{1}{4c}$; $\frac{1}{3d^2} - \frac{1}{2d}$ (3) $\frac{1}{p + 1} - \frac{1}{2p + 3}$;

$$\frac{1}{q+2} - \frac{2}{q+4}; \quad \frac{5}{r+2} - \frac{4}{r+3}; \quad \frac{1}{s-4} - \frac{1}{s+5}.$$

5 Express as single fractions: (a) $\dfrac{x^2+3x}{x^2-9} \div \dfrac{x^2}{4}$

(b) $\dfrac{6x^2-6}{x} \div \dfrac{3x+3}{x^2}$ (c) $\dfrac{1}{x^2} - \dfrac{1}{x^2+x}$

(d) $\dfrac{1}{x^2+3x+2} + \dfrac{1}{x+2}$.

6 Find the value of $\dfrac{x^2}{x^2-x}$ when $x = 2$.

Simplify $\dfrac{x^2}{x^2-x}$ and check that, when $x = 2$, your answer gives the same value as before.

7 Repeat question 6 for (a) $\dfrac{x^2-1}{x+1}$ (b) $\dfrac{2x-6}{x^2-3x}$

(c) $\dfrac{x^2+3x}{x^2+7x+12}$ (d) $\dfrac{1}{x} - \dfrac{1}{x^2+x}$ (e) $\dfrac{1}{2x-3} + \dfrac{1}{2x+3}$

(f) $\dfrac{1}{x+3} + \dfrac{4}{x^2+2x-3}$

8 $p = 3$, $q = -2$ and $r = -4$. Find the value of:
$p+q+r$; $p-q-r$; pqr; pq^2; $(pq)^2$; p^2+q^2; $(p+q)^2$; $p^2+2pq+q^2$; p^2-q^2; $(p-q)^2$; $(p+q)(p-q)$; $\dfrac{pq}{r}$; $\dfrac{1}{p}+\dfrac{1}{q}$; $\dfrac{1}{p+q}$; $\dfrac{p+q}{pq}$.

9 Arrange $2x^2$, $(2x)^2$, $1/2x$ and $1/x^2$ in order of increasing size when (a) $x = 3$, (b) $x = \frac{1}{2}$, (c) $x = -2$.

10 Change the subject of the following formulae to the symbol indicated.

$F = \dfrac{mv^2}{r}$, subject r,m,v; $v = \sqrt{2gh}$, subject g,h;

$I = \dfrac{C}{d^2}$, subject C,d; $p = q(r+s)$, subject; q,r,s

$r = \sqrt{\dfrac{V}{\pi h}}$, subject h,V; $t = 2\pi\sqrt{\dfrac{I}{MH}}$, subject I,M,H

11 Express p in terms of q if: (a) $q = \dfrac{2p-3}{p}$

(b) $q = \dfrac{3p + 5}{4p}$ (c) $q = \dfrac{5p - 2}{p + 4}$ (d) $q = \dfrac{3p^2 + 2}{p^2 - 1}$

(e) $q = \dfrac{4 - p^2}{4 + p^2}$.

12 $y = x^2 - 7x + 10$. (a) Find y when $x = 3, \frac{1}{2}$ and -2. (b) Find x when $y = 0$ and 4.

13 $y = 2x^2 - 5x - 3$. (a) Find y when $x = 2, -2$ and $\frac{3}{4}$. (b) Find x when $y = 0$ and 4.

14 Find x if $3x + 9 = 15$; $4x - 3 = 13$; $\frac{1}{2}(3x + 1) = 8$; $\frac{3}{4}(x - 5) = 9$; $\frac{1}{3}(2x - 3) = (x + 4)$; $x + \frac{1}{2}(x + 3) = 6$; $\frac{1}{2}(3x + 2) - \frac{3}{4}(x - 2) = 7$; $\frac{1}{4}(x + 1) - \frac{1}{3}(2x - 1) + 4 = 0$.

15 State whether the following simultaneous equations have (a) one pair of solutions, (b) no solutions, (c) an unlimited number of solutions.

(1) $2x + y = 6$, $3x - y = 24$ (2) $3r - 2s = 4$, $4r + 2s = 38$ (3) $2x + 5y = 26$, $3x - 2y = 1$

(4) $12x + 8y = 14$, $4y + 6x = 7$ (5) $3x - 2y = 10$, $2x + y = 9$ (6) $4x + 10y = 6$, $2x + 5y = 4$.

Solve those in category (a).

16 What is the value of z if (a) $z(z + 5) = 0$, (b) $(z + 2)(z - 3) = 0$, (c) $(2z + 1)(3z - 2) = 0$, (d) $(z + 1)(z + 2)(z - 3) = 0$?

17 Find the solution sets of: (a) $\{x: x^2 + x - 6\} = 0$, (b) $\{y: y^2 + 7y + 10 = 0\}$, (c) $\{z: 3z^2 = 11z + 4\}$, (d) $\{p: 6p^2 + 11p = 10\}$, (e) $\{q: 5q^2 = 14q + 3\}$, (f) $\{r: 4r^2 = 15 - 17r\}$, (g) $\{s: 4s^2 = 7s + 15\}$.

18 For what values of x – if any – does (a) $x(x - 3) = 0$, (b) $x(x + 7) = x^2 + 7$, (c) $x(x - 3) = x^2 - 3x$, (d) $x^2 + 4x - 5 = 0$?

19 Solve, giving answers correct to 2 decimal places: (a) $x^2 - 5x + 1 = 0$, (b) $x^2 + 7x + 3 = 0$, (c) $2x^2 + 9x + 1 = 0$, (d) $2x^2 - 3x - 7 = 0$, (e) $5x^2 + 3x = 3$, (f) $6x^2 + x = 4$.

20 Calculate y to three significant figures if: (a) $2y^2 - 3y - 1 = 0$, (b) $5y^2 - 4y - 3 = 0$, (c) $3y^2 = 1 - 5y$, (d) $4y^2 = 3y + 6$.

21 Form quadratic equations with solutions (a) $x = \frac{1}{2}(1 \pm \sqrt{5})$, (b) $y = \frac{1}{2}(3 \pm \sqrt{21})$, (c) $s = \frac{1}{4}(-2 \pm \sqrt{28})$, (d) $t = \frac{1}{8}(5 \pm \sqrt{73})$.

22 $P = 6x + 3$ and $Q = 2x - 5$. Find (a) $P + Q$, (b) $P - 2Q$, (c) PQ, (d) Q^2, (e) x, if $P = 2Q$, (f) x, if $P = Q^2$.

23 $L = 3x + 5$ and $M = 4 - 2x$. Find (a) $2L + 3M$, (b) $L - M$, (c) x in terms of L, (d) M in terms of L, (e) x, if $L + M = 10$, (f) x, if $L = 4/M$.

24 $P = 3x^2$, $Q = 6x$ and $R = 4x + 3$. (a) Find the values of P, Q and R when $x = 3, \frac{1}{3}$ and -4. (b) Find x if (1) $Q =$ R, (2) $P = Q$ and (3) $P = R + 1$. (c) Express x in terms of R. (d) Express P in terms of Q. (e) Write as single fractions $\frac{1}{P} + \frac{1}{Q}$ and $\frac{3}{Q} - \frac{1}{R}$.

25 $R = x^2 + 4x$, $S = x^2 + 3x - 4$ and $T = x^2 - 3x + 2$. (a) Factorise R, S and T. (b) Find x if $R = 0$. (c) Find x if $S = T$. (d) Find x if $S = 2T$. (e) Simplify $\frac{RT}{S}$. (f) Simplify $\frac{1}{R} - \frac{1}{S}$.

26 $R = 4x + 5$ and $S = 2x - 3$. Find x if (a) $R = 9$, (b) $S = -5$, (c) $R + S = 16$, (d) $R - S = 4$, (e) $R/S = 6$.
Find the two values of x if (a) $R = x^2$, (b) $S^2 = 1$, (c) $RS + 15 = 0$.

27 Write down the next two terms and an expression for the nth term in the sequences:
(a) 2, 4, 6, 8. . ., (b) 1, 5, 9, 13. . ., (c) 1·3, 2·4, 3·5, 4·6. . ., (d) $\frac{1}{2}$, $\frac{2}{3}$, $\frac{3}{4}$, $\frac{4}{5}$. . ., (e) $2x$, $3x^2$, $4x^3$, $5x^4$. . ., (f) 1, −1, 1, −1. . .

28 Say whether the following statements are true for (1) all values of x, (2) only two values of x, (3) only one value of x, (4) no values of x.
(a) $3x + 2 = 11$, (b) $4(x - 3) = 4x - 12$, (c) $\frac{1}{2}(x + 3) = x + 5$, (d) $x(x + 2) = 0$, (e) $(x + 1)^2 = x^2 + 2x + 1$, (f) $(x + 1)^2 = x^2 + 1$.

29 Replace the symbol 'o' by $=, -, \times, +$ or $\div$ if $\frac{1}{x} \text{ o } \frac{1}{2x}$ equals (a) $\frac{1}{2x}$, (b) $\frac{1}{2x^2}$, (c) 2, (d) $\frac{3}{2x}$.

30 Repeat question 29 for $\frac{1}{x} \text{ o } \frac{1}{x^2}$ equals (a) $\frac{1}{x^3}$, (b) $\frac{x + 1}{x^2}$, (c) x, (d) $\frac{x - 1}{x^2}$.

31 Find the angles of a quadrilateral if (a) the angles are $x°$, $(x + 20)°$, $(x + 30)°$ and $(x + 50)°$, (b) the angles are $x°$, $2x°$, $3x°$ and $(3x - 72)°$, (c) three of the angles are equal and the fourth is 52° larger than each of these, (d) two of the angles are equal, a third is 20° larger and the fourth 40° larger than these.

32 Joy has £x, Roy has £3 more than this. Write in terms of x (a) the amount Roy has, (b) the amount each has if Roy gives Joy £5.

Find x if Joy now has twice as much as Roy.

33

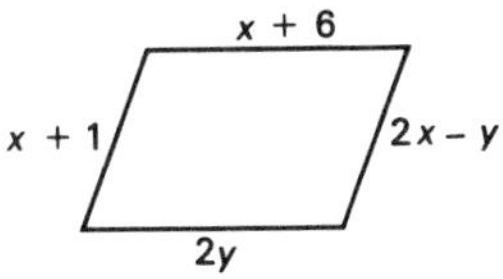

Fig. 4.1

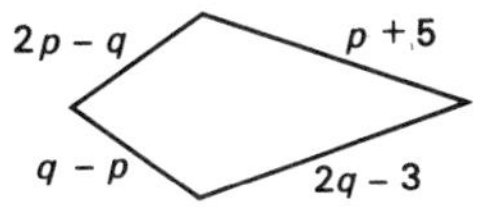

Fig. 4.2

(a) Calculate x and y if figure 4.1 is a parallelogram.
(b) Calculate p and q if figure 4.2 is a kite.

34 State the value of K if the following are perfect squares. (a) $x^2 + 6x + K$, (b) $y^2 + 16y + K$, (c) $p^2 - 10p + K$, (d) $4r^2 + 24r + K$, (e) $25h^2 - 60h + K$.

35 A right angled triangle has sides x, $2x - 1$ and $2x + 1$. (a) which of these is the hypotenuse? (b) Express in terms of x (1) the perimeter, (2) the area. (c) Calculate the value of x.

36 A rectangle has sides x and $(2x + 3)$. Find x if (a) one side is 10 cm longer than the other, (b) the perimeter of the rectangle is 96 cm, (c) the ratio of the sides is 5:2, (d) the area of the rectangle is 35 cm².

37 A model rocket rises to a height h metres in a time t seconds where $h = 100t - 5t^2$. Find (a) h when $t = 6$, 10 and 25 s, (b) t when $h = 375$, 480 and 0 m, (c) t – to the nearest tenth – when $h = 400$ m.

38

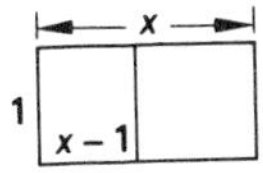

Fig. 4.3

Figure 4.3 shows a 'Golden Mean' rectangle, which has the property that $x:1 = 1:x - 1$.

Form an equation in x and calculate the Golden Mean to two decimal places.

Relations

Exercise 5

1 D = {d, a, y} and N = {n, i, g, h, t}. How many elements are there in the product sets DN, ND, DD and NN? Which of these products sets contains the ordered pairs (d,g) (y,d) (i,a) (h,h) (t,g) (a,y) and (t,t)?

How many possible relations can be formed from the elements of the product set DN?

2 Write out the product set PQ where P = {3, 6, 9} and Q = {−, +, ×, 0}. Give an example of a many-many, one-many, many-one and one-one relation formed from PQ.

3 A = {1, 3, 5} and B = {2, 4, 6, 8}. Tabulate the product set AB. List the ordered pairs from AB for the relations (a) . . . is one less than . . ., (b) . . . is a factor of . . ., (c) . . . differs by 3 or more from . . ., (d) . . . leaves a remainder 2 when divided by . . ., (e) . . . leaves a remainder 1 when divided into . . ..

4 P, Q and R are sets and $n(\mathrm{P}) = 5$, $n(\mathrm{Q}) = 6$ and $n(\mathrm{R}) = 12$. How many ordered pairs are there in the product sets PQ, PR, QR, RQ, PP and RR?

What can be said about Q and R if (a) $n(\mathrm{PQ}) = n(\mathrm{PR})$, (b) QR = RQ?

5 Make a table of the product set AB where A = {1, 6, 12} and B = {2, 4, 12, 24}. How many elements of the product set AB satisfy the relation (a) . . . is equal to . . ., (b) . . . is half of . . ., (c) . . . is greater than . . ., (d) . . . is a factor of . . ., (e) . . . differs by 4 or less from . . . ?

Which of these relations are functions? Classify the above relations as many-many etc.

6 Give three ordered pairs from each of the following functions (a) $f{:}x \rightarrow 4x$, (b) $f{:}x \rightarrow 2x + 1$, (c) $f{:}x \rightarrow x^3$, (d) $f{:}x \rightarrow \dfrac{x+3}{5}$.

State the inverses of the above.

7 Give three ordered pairs from each of the following relations. Take x as a whole number.

(a) $\{(x,y){:}y = \tfrac{1}{2}x\}$, (b) $\{(x,y){:}y = 2x^2\}$, (c) $\{(x,y){:}2x + y \leq 10\}$, (d) $\{(x,y){:}y$ is a multiple of $x\}$, (e) $\{(x,y){:}x < y\}$, (f) $\{(x,y){:}xy = 12\}$, (g) $\{(x,y){:}y$ is the next prime number after $x\}$.

8 Repeat question 7 taking x as a negative integer.

9 Given $f:x \to 3x + 1$ and $g:x \to \dfrac{12}{x}$ find: $f(3)$ $f(-3)$ $f(-\frac{1}{3})$ $g(4)$ $g(6)$ $g(\frac{1}{2})$ and $g(-\frac{1}{8})$.
Find x if $f(x) = 25$, $g(x) = 72$. Calculate $g[f(5)]$ and $f[g(5)]$. Express in simplest form $g[f(x)]$ and $f[g(x)]$.

10 $f(x) = (3x + 2)/4$ and $g(x) = 2/x$. Find $f(2)$ $f(6)$ $f(-6)$ $f(\frac{1}{2})$ $f(-\frac{1}{2})$ $g(4)$ $g(-4)$ $g(\frac{1}{4})$ $f[g(4)]$ $g[f(4)]$.
Express in simplest form $f[g(x)]$ $g[f(x)]$ $g[g(x)]$ and $f[f(x)]$.

11 $f(x) = 2x + 1$, $g(x) = 2x^2$ and $h(x) = 4/x$.
Calculate the images $f(5)$ $g(6)$ $g(-6)$ $f(-\frac{1}{2})$ $h(\frac{1}{2})$ $h(12)$ $g[f(2)]$ $f[g(2)]$ $g[h(4)]$ $h[g(4)]$ $f[h(\frac{1}{3})]$ $h[f(\frac{1}{3})]$.
Write in simplest form $g[f(x)]$ $f[g(x)]$ $g[h(x)]$ $h[h(x)]$ $f\{g[h(x)]\}$ $h\{g[f(x)]\}$.

12 Find the range of the following functions.

Domain	Reals	Whole numbers	Multiples of 6
Function	$x \to x^2$	$x \to 2x$	$x \to 5x$
Domain	Integers	Integers	Whole numbers
Function	$x \to 2x - 1$	$x \to 2x^2$	$x \to \dfrac{x}{2}(x + 1)$
Domain	Reals from -6 to $+6$		
Function	$x \to x^2$		

13 Given $f:x \to x^2$, find $f(0)$ $f(1)$ $f(2)$ $f(3)$ $f(-1)$ $f(-2)$ and $f(-3)$. Use these values to sketch the graph of the function for x a real number.

14 For $f:x \to 2/x$ calculate $f(4)$ $f(2)$ $f(1)$ $f(-1)$ $f(-2)$ and $f(-4)$. Why is $f(0)$ omitted? Assuming x is a real number, sketch a graph of this function.

15 Calculate $f(3)$ $f(2)$ $f(1)$ $f(0)$ $f(-1)$ $f(-2)$ and $f(-3)$ for $f:x \to 2x^2 - 3$. Use these values to sketch a graph of the function for x real. On your sketch shade $\{(x,y):y \geqslant 2x^2 - 3\}$.

16 Sketch a graph of $\{(x,y):y = 2x - 3\}$. On your graph shade $\{(x,y):y \leqslant 2x - 3\}$. Take real values of x.

17 If $f(x) = 2^x$, calculate $f(2)$ $f(1)$ $f(0)$ $f(-1)$ and $f(-2)$. Use these values to sketch a graph of the function, for real values of x.

18 Taking x as real, sketch graphs of:
(a) $y = x$; $y = 3x$; $y = 3x + 4$. (b) $y = x^2$; $y = x^2 - 5$; $y = x^2 + 5$; $y = 5 - x^2$. (c) $y = 2x$; $y = 2 + x$; $y = 2/x$.

19 Draw a graph to illustrate the relation R = 'has a factor in common with . . .' on {3, 4, 5, 6}.

Explain how your graph shows that R is reflexive and symmetric. Is this an equivalence relation? Say why.

20 Draw a graph to illustrate $x\text{R}y$ on the set {4, 8, 14, 16} where R = '. . . leaves the same remainder when divided by 4 as . . .'. Explain how your graph can be used to test whether R is reflexive or symmetric. Is R an equivalence relation?

21 Say whether the following relations between triangles are R, reflective, S, symmetric, T, transitive. (a) . . . is similar to . . ., (b) . . . is congruent to . . ., (c) . . . is the mirror image of . . ., (d) . . . is an enlargement of . . ., (e) . . . has the same area as . . ., (f) . . . is not similar to . . ., (g) . . . has less area than . . ..

22 Which of the relations in question 21 are equivalence relations? Which relations order the set?

23 Classify the following relations between a collection of sets as reflexive(R), symmetric(S) or transitive(T). (a) . . . is a proper subset of . . . (b) . . . is the complement of . . . (c) . . . has the same number of elements as . . . (d) . . . has less elements than . . .

Which relations order the sets? Which are equivalence relations?

24 Classify the following relations (R) reflexive, (S) symmetric and (T) transitive.

(a) . . . is in the same house as . . . for a set of 5th Formers.

(b) . . . owns the same make of motor cycle as . . . for 6th Form motor cyclists.

(c) . . . is faster over 100 m than . . . for a set of sprinters.

(d) . . . is next to . . . for houses in a row.

(e) . . . is married to . . . for inhibitants of a village. No bigamy!

25 Which relations in question 24 (1) partition the sets, (2) order the sets? Which are equivalence relations?

26 Say which of the following are equivalence relations and which are ordering relations.

(a) . . . lies to the North of . . . for towns in this country.

(b) . . . is a proper or improper subset of . . . for a collection of subsets of a set.

(c) . . . has the same shape as . . . for a set of quadrilaterals.

(d) . . . has more jam than . . . for a set of doughnuts.

(e) . . . is browner than . . . for eggs in a box.

Variation

Exercise 6

1 Express in words and in the form $y = k.f(x)$: $y \propto x$; $y \propto 1/x$; $y \propto 1/x^2$; $y \propto x^3$; $y \propto \sqrt{x}$.

2 Sketch graphs to illustrate the relations in question 1 for non-negative values of x.

3 Given that $y \propto x^2$, (a) express the relation in words (b) draw a sketch graph of the relation for $x \geqslant 0$ (c) say if this is an example of direct or inverse variation. (d) If x is doubled what happens to y? (e) If x increases by 20% what is the percentage change in y? (f) Given that $y = 6$ when $x = 3$, find a formula connecting y and x and the value of y when $x = 12$.

4 Repeat question 3 for (1) $y \propto 1/x$, (2) $y \propto x^3$.

5 'y varies directly as the square root of x.' (a) Express this statement in symbols. (b) Sketch a graph of this form of variation. (c) What change in x is needed to increase y three times? (d) Find a formula connecting y and x if $y = 8$ when $x = 16$. (e) Use this formula to find y when $x = 81$ and x when $y = 72$.

6 The time of swing 't' of a pendulum is proportional to the square root of the length 'l' of the pendulum. (a) Express this statement in symbols. (b) In order to halve the time of swing what change is needed in the length? (c) Find a formula connecting t and l if $t = 3$s when $l = 2\frac{1}{4}$ m. (d) Use this formula to find t when $l = 3{\cdot}24$ m and l when $t = 4$s. (e) What is the change per cent in the time of swing if the length is increased 44%?

7 The distance 's' fallen by a body is directly proportional to the square of the time of descent 't'. (a) Express this statement in symbols. (b) Find a formula connecting s and t if $s = 20$ m when $t = 2$ s. (c) Use your formula to find s when $t = 5$ s and t when $s = 180$ m. (d) Express the relation between t and s in symbols.

8 State the form of variation you would expect to find between the following: (a) 'A', the surface area of a saucer and 'r', its radius. (b) 'C', the circumference of a circle and 'r', its radius. (c) 'V', the volume of a cube and 'l', the length of its edge. (*d*) 'N' the number of lines ruled on a page and 'd' the distance

between the lines – give two answers, one for exercise paper the other for squared paper.

9 What forms of variation would you expect between the following? (a) 'T' the time of a journey and 'v' the average speed. (b) 'N' the number of slices from a ham and 't' the thickness of a slice. (c) 'L' the length of a diagonal of a square and 'A' its area. (d) 'S' the number of squares that can be cut from a sheet of paper and 'l' the length of a side of a square.

10 The intensity of illumination 'I' from a light bulb varies inversely as the square of the distance 'd' from the bulb. (a) As d increases what happens to I? (b) Sketch a graph to illustrate this form of variation. (c) If d is halved what is the change in I? (d) Calculate the percentage change in I for a 50% increase in d – answer to the nearest whole number.

11 The number of ants 'N' in an anthill is directly proportional to the cube of the height 'h' of the anthill. (a) Express this relation in symbols. (b) Sketch a graph to illustrate this relation. (c) If there are 10 000 ants in a hill 20 cm high how many are there in a hill (1) 40, (2) 30 and (3) 50 cm high?

Inequalities

Exercise 7

1 Complete the following by inserting the signs $>$, $<$ or $=$. (a) 55 . . . 56, (b) 1/55 . . . 1/56, (c) 8 . . . 12, (d) -8 . . . -12, (e) 1·6 . . . 8/5, (f) $\frac{2}{3} \ldots \frac{3}{2}$, (g) 2/7 . . . 0·29, (h) $\pi \ldots 3{\cdot}14$, (i) $1 \ldots 1\frac{3}{4} \ldots 2$, (j) $-3 \ldots -2\frac{1}{2} \ldots -2$.

2 If $m > n$ what can be said about (a) $m + 6$ and $n + 6$, (b) $m - 3$ and $n - 3$, (c) $5m$ and $5n$, (d) $-5m$ and $-5n$, (e) $\frac{1}{4}m$ and $\frac{1}{4}n$, (f) m^2 and n^2, (g) m and $-n$, (h) $-m$ and $-n$?

3 Complete the following by inserting the signs $>$, $<$, $\Rightarrow$ and $\Leftrightarrow$. (a) $5 > 3 \ldots 3 \ldots 5$, (b) $19 \ldots 21 \ldots 21 > 19$, (c) $p > q \ldots \frac{1}{2}p \ldots \frac{1}{2}q$, (d) $p < q \ldots p + r \ldots q + r$, (e) $r < t \ldots r^2 \ldots t^2$, (f) $s > t \ldots -s \ldots -t$, (g) $p > q$ and $q > r \ldots p \ldots r$.

4 Deduce, where possible, the consequences of the following pairs of statements. (a) $n > p$ and $p > q$ (b) $n \geqslant 7$ and $p = n$ (c) $n < 5$ and $p \leqslant n$ (d) $n \leqslant 6$ and $p \geqslant 6$ (e) $n < p$ and $nq < pq$.

5 Solve the following inequations for $x \in R$.
(a) $3x > 6$, (b) $4x + 5 \leqslant 21$, (c) $3(x + 4) < 27$,
(d) $4x + 2 \leqslant x + 11$, (e) $x + 2 \leqslant 4x + 14$,
(f) $2(x + 3) \geqslant 3(x - 7)$.

6 For $y \in R$, find the range of values of y for which
(a) $5y \leqslant 9$, (b) $2y - 4 \geqslant 7$, (c) $4(y - 5) < 6$,
(d) $\frac{1}{2}(3y - 4) \leqslant 6$, (e) $\frac{3}{4}(y + 5) > 8$, (f) $\frac{3}{8}(y - 3) \leqslant 7$.

7 Solve $2x + 3 \leqslant 17$ for (a) x a real number, (b) x a positive multiple of 3.

8 Solve $3x + 5 \leqslant 26$ for (a) x a real number, (b) x a whole number, (c) x a positive even number.

9 Solve $\frac{1}{3}(x - 5) \leqslant 4$ for (a) x a real number, (b) x a positive multiple of 5, (c) x a positive prime number.

10 Solve $4x - 7 \geqslant 17$ for (a) x a real number, (b) x a factor of 30, (c) x a factor of 48.

11 Solve $5 < 2x - 3 < 16$ for (a) x a real number, (b) x an odd number, (c) x a multiple of 3.

12 Solve $2 \leqslant \frac{1}{3}(x + 1) \leqslant 8$ for (a) x a real number, (b) x a prime number, (c) x a multiple of 6.

13 Find the range of values of y that satisfy the following inequations for $y \in R$.
(a) $-3 < y + 3 < 7$, (b) $-5 < 2y + 3 < 9$,
(c) $-3 \leqslant 2(y + 3) \leqslant 7$, (d) $0 \leqslant \frac{1}{2}(y + 3) \leqslant 3$,
(e) $-6 < \frac{1}{3}(y - 5) < 6$.

14 (a) Choose three values of x that show that $1/x < 4$ does not always imply that $1 < 4x$.
(b) Solve $1/x < 4$, considering the cases $x > 0$ and $x < 0$.
(c) Solve $12/x < 3$; $15/2x < 5$; $18/5x < 6$; $18/5x > 6$.

15 Solve the following inequations for $x \in R^+$.
(a) $6/x > 5/9$, (b) $6 + 1/x \geqslant 15$, (c) $5/x - 3 \leqslant 10$.
What changes are there in the above answers if $x \in R^-$?

16 Sketch graphs and outline the region in which
(a) $x > 0$ and $y \leqslant 4$, (b) $x \leqslant 0$ and $0 \leqslant y < 6$,
(c) $-2 < x < 2$ and $-3 < y < 3$, (d) $x \geqslant 0$, $y \geqslant 0$ and $x - y \geqslant 0$, (e) $x \leqslant 0$, $y \leqslant 0$ and $x - y \leqslant 0$,
(f) $x \leqslant 0$, $y \geqslant 0$ and $x + y \geqslant 0$, (g) $x \geqslant 0$, $y \leqslant 0$ and $x + y \leqslant 0$.

17 Sketch graphs of $x = 4$, $y = 3$ and $2x + y = 6$. Shade in different colours the regions that satisfy the conditions
(a) $y \geqslant 3$ and $2x + y \leqslant 6$, (b) $x \geqslant 4$, $y \leqslant 3$ and $2x + y \geqslant 6$,

(c) $x \geqslant 0$, $0 \leqslant y \leqslant 4$ and $2x + y \leqslant 6$, (d) $x \leqslant 4$, $y \leqslant 0$ $2x + y \leqslant 6$.

18 A coach carries up to 48 passengers and costs £16 per day to hire. State conditions for 'n', the number of passengers on a day trip, if the maximum fare is 50p per passenger.

19 On a 126 km journey speeds are to be kept below 72 km/h and the journey is to be completed in $2\frac{1}{4}$ h, at the most. State conditions for the speed v.

20 A heavy lorry has to cross three bridges and travel through two tunnels. The bridges are 5, 6 and 4·5 metres wide and carry loads of 12, 10 and 13 tonnes. The tunnels are 5·6 m high and 5·5 m wide and 6·4 m high and 4·4 m wide. What conditions of height, width and weight must the lorry and its load satisfy?

21 The sides of a rectangle are measured as 20 cm $\pm$ 0·5 cm and 15 cm $\pm$ 0·5 cm. Calculate the range of possible values of (a) the perimeter, (b) the area of the rectangle.

22 John Vague said he bought six or seven boxes of chocolates and he thought there were between fifteen and twenty chocolates in each, of eight to ten varieties. Denote the varieties by 'v' and the total number of chocolates by 'n' and give the range of values of v and n.

Graphical Work

Exercise 8

1 Draw a graph of $y = \dfrac{90}{x}$, using $x = 10, 15, 20, 30, 45$ and 60.

Take 1 cm to 5 units for the x-axis and 2 cm to 1 unit for the y-axis. Use your graph to find:

(a) the width of a rectangle with area 90 cm², length 25 cm,
(b) the height of a triangle with area 45 cm², and base 18 cm,
(c) the time of a journey of 90 km at 42 km/h,
(d) the average speed of a car that completes a 90 km journey in 1 h 48 min.

2 Plot a graph of $y = \dfrac{12}{x}$, using $x = \frac{1}{2}$, 1, 2, 3, 4, 5, and 6.

Take 1 cm to 2 units for the y-axis and 2 cm to 1 unit for the x-axis. Use your graph to find: (a) $\dfrac{12}{1{\cdot}4}$, $\dfrac{12}{2{\cdot}4}$, $\dfrac{12}{4{\cdot}4}$, (b) the speed of a pigeon that flies 12 metres in 1·8 seconds, (c) the gradient of the graph at $x = 1{\cdot}5$, (d) for what value of x the gradient is -3 units.

3 Using $x = \frac{1}{5}, \frac{1}{2}$, 1, 2, 3, 4 and 5, plot a graph of $y = \dfrac{1}{x}$. Choose scales to make the graph as large as possible.

Find the gradient of the graph at $x = 1$. What would this gradient give if y represented distance and x time?

4 Make another copy of the graph in question 3 and on the same axes plot the straight line $y = x + 2$. Read off the value of x where the two graphs cut. Explain why this is one solution of $x^2 + 2x - 1 = 0$. How would you find the other solution?

5 A jet transport moves along the runway and its speeds in m/s at intervals of 10 s until it becomes airborne are recorded as:

Time	s	0	10	20	30	40	50	60	70	80
Speed	m/s	0	2	7	16	28	42	53	59	61

Plot a speed-time graph for this take-off run using a scale of 1 cm to 5 m/s vertically and 1 cm to 5 s horizontally. From your graph estimate: (a) the speeds at 25, 35 and 45 s, (b) the maximum acceleration, (c) the length of the take-off run.

6 The rate of flow of oil from a tank is measured in l/min at intervals of 2 min.

Time min	0	2	4	6	8	10	12	14	16	18
Flow l/min	24·5	16·5	12	9	7	5	4	3	2	1

Plot a rate of flow – time graph using 2 cm to 5 l/min vertically and 1 cm to 1 min horizontally. Use your graph to find, (a) the rates of flow at 5, 7 and 9 min, (b) when the rate of flow fell below 15 l/min, (c) the total flow in the 18 min.

7 The rate of climb of a prototype rocket plane in its first 80 s of flight was recorded as:

Time s	0	10	20	30	40	50	60	70	80
Climb m/s	0	60	170	330	440	490	400	270	190.

Plot a rate of climb/time graph using 1 cm to 25 m/s vertically and 1 cm to 5 s horizontally. Estimate (a) the rates of climb at 25, 35 and 65 s, (b) the time for which the rate of climb exceeded 350 m/s, (c) the total gain in height in the 80 s.

8 Calculate the missing items in the table of values for $y = 3^x$

$x =$	-2	-1	0	1	2	2·5	3
$y =$		$\frac{1}{3}$		3	9	15·5	

Take scales of 2 cm to 1 unit for the x-axis and 1 cm to 2 units for the y-axis and plot the graph of $y = 3^x$.

Use your graph to estimate (a) the values of $3^{0{\cdot}5}$, $3^{1{\cdot}5}$, (b) the values of x if $3^x = 8$, $3^x = 18$, (c) the power to which 3 must be raised to give 20 ($\log_3 20$).

9 The number of victims of a 'flu epidemic is found to be increasing 3 times every 5 days. Use your graph from question 8 to estimate (a) how many times more victims there are in 6 days, 8 days and 2 weeks, (b) the number of days before the total victims increases 10 times.

10 The penguin population on an island in Antarctica was recorded at intervals of 5 years, except in 1945.

Year	1925	'30	'35	'40	'50	'55	'60	'65
Penguins	830	610	460	370	250	220	200	180

Plot a population-time graph using 1 cm to 50 penguins vertically and 1 cm to 5 years horizontally. Estimate (a) the 1945 population, (b) the population in 1938 and 1958, (c) the rate of fall of population in 1930, 1940 and 1950. Is the rate of fall increasing or decreasing?

11 Draw an x and a y axis and mark these with scales of 1 cm to 1 unit. Outline the region satisfying the following conditions: $4x + 3y \leqslant 60$, $4x + 3y \geqslant 12$, $0 \leqslant y \leqslant 12$, $0 \leqslant x \leqslant 10$.

Make a table of the possible pairs of values of x and y if x and y are multiples of 4. Which of these pairs gives a maximum value for (a) $2x + y$, (b) $x + 2y$?

12 Repeat question 11 for x and y multiples of 3.

13 A photographic society plan to spend between £12 and £60 on lighting equipment. They intend to purchase two types of floodlight; type X costing £4 each and type Y costing £3 each. They require 3 of type X to every 5 of type Y. Use your graph from question 11 to find their maximum and minimum costs.

14 Repeat question 13 for X and Y in ratio 1:2.

15 Using 1 cm to 1 unit on both x and y axes, outline the region for which $x \geqslant 0$, $y \geqslant 0$, $x \leqslant 5$, $y \leqslant 8$, $3x + 2y \leqslant 24$. Find the maximum value of $4x + 3y$ if x and y are whole numbers.

What is the maximum value of $2x + 3y$ under these conditions?

16 The owner of a riding school keeps horses and ponies. The upkeep for a horse is £1.50 per week and that for a pony £1.00 per week. He limits his weekly expenses to £12. There are sufficient stables for up to 5 horses and up to 8 ponies but neither can use the others' stables because of size and draught.

How many of each should he keep to maximise a profit of £2 per horse and £1.50 per pony?

17 What alternative solutions does the owner in question 16 have if his profit changes to £2.00, for a horse and £2.00 for a pony?

18 With scales of 1 cm to 1 unit on both axes, outline the region satisfying $x \geqslant 2$, $y \geqslant 2$, $5x + 7y \leqslant 70$, $3x + 2y \leqslant 30$. List the solution set if x and y are even numbers. Which pairs of values maximise (1) $3x + 2y$, (2) $2x + 3y$?

19 Repeat question 18 if (a) x and y are both odd numbers, (b) both are multiples of 3.

20 A bird fancier keeps budgerigars (X) and canaries (Y) in breeding pairs. Budgies cost him £2.50 each and canaries £3.50 each. He has £35 to invest in new stock and decides to keep at least one pair of each.

On the average, budgies produce 3 young and canaries 2 young but he has nursery space for only 15 young. He sells the young birds for £2 each. Show that his profit is £$(3x + 2y)$ and find the numbers he should keep to maximise this profit.

21 120 scouts set off for camp, loaded with 3 600 kg of equipment. They are transported in 2 types of truck:

	No	Holds	Kit capacity	Cost
Truck X	8	12 scouts	400 kg	£4
Truck Y	6	15 scouts	300 kg	£5

(a) Find the arrangement that gives the least cost. (b) Find the cost of the three cheapest arrangements.

22 Aunt Matilda has a lawn and £4.80 to spend. She is very fond of plastic gnomes (x) and plastic dwarfs (y). Gnomes cost 60p and dwarfs 40p. Show that $3x + 2y \leqslant 24$.

She feels that she must have at least 2 gnomes and at least 3 dwarfs but certainly not more than 7 dwarfs. Plot a graph to illustrate these restrictions, using scales of 1 cm to 1 creature.

From your graph find (a) the total number of possible solutions, (b) the solutions that use all the money, (c) the maximum number of creatures Auntie can buy, (d) the cheapest way of doing this.

Two-Dimensional Figures

Exercise 9

Use $\pi = 22/7$ or $3{\cdot}14$, as appropriate.

1 Calculate the lengths of plastic required to edge the following tables. (a) A rectangular table 1·8 m by 0·7 m. (b) A square table of side 1·6 m. (c) A circular table of radius 0·98 m. (d) A semi-circular table of radius 0·98 m. (e) A hexagonal table of side 0·88 m. (f) A semi-hexagonal table of side 0·88 m.

2 Calculate the area of (a) a square of side 10 cm, (b) a circle of diameter 10 cm, (c) a circle of radius 10 cm, (d) an equilateral triangle of side 10 cm, (e) a regular hexagon of side 10 cm.

3 Each of the following figures has an area of 1 hectare. Calculate in metres (a) the side, if the figure is a square, (b) the width, if the figure is a rectangle of length 125 m, (c) the base, if the figure is a triangle of height 240 m, (d) the radius, if the figure is a circle, (e) the edge, if the figure is an equilateral triangle.

These questions refer to Fig. 9.1, in which the lengths are in cm.

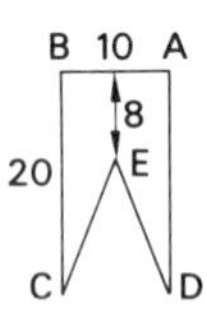

Fig. 9.1

4 Calculate (a) the area of the figure, (b) its perimeter, (c) the distance AE.

5 Two such figures are placed with their corners C and their corners D touching. Calculate the area enclosed between them.

6 Make a one-quarter full size drawing of the figure. Mark any axes or centres of symmetry on your drawing.

Sketch how this shape could be used to tessellate an area.

7 One such figure slides 4 cm in the direction AD. Calculate the area that (a) AB and (b) CED slides over. Account for the relation between the two answers.

8 One such figure is rotated in its own plane through 360° about the vertex A. State, as a multiple of π, the area swept out by AB, AD, AC and BC.

9 Figure 9.2 shows a diamond shape cut from a rectangle of sides 18 and 6 cm. A second diamond, similar to the first with width 2 cm, is cut from the centre.

Calculate the shaded area that remains.

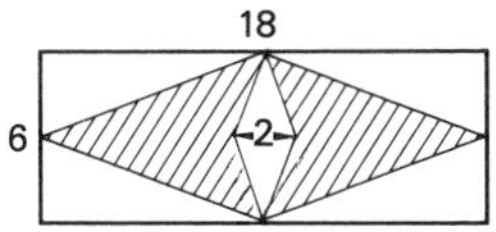

Fig. 9.2

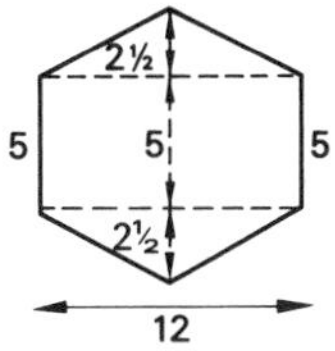

Fig. 9.3

10 Refer to Fig. 9.3, for these questions. The lengths are in cm.

(a) Make a half-full-size drawing of the figure. (b) Mark on your drawing any axes or centres of symmetry. (c) Make a sketch to show how such shapes could be fitted together to cover an area. (d) Calculate the area of the figure. (e) Calculate its perimeter.

11 Calculate the area a line LM, 7 cm long, passes over if (a) it is translated 12 cm at 90° to LM, (b) if it is translated 10 cm at 30° to LM, (c) if it is rotated through 360° about L.

What area is swept over twice, if it is rotated 360° about a point 2·1 cm from L?

12 A kite WXYZ has sides WX = WZ = 10 cm, YX = YZ = 17 cm and XZ = 16 cm. (a) Make a one-quarter size drawing of the kite. (b) Name three pairs of congruent triangles in the figure, together with its two diagonals. (c) Calculate the length of the diagonal WY. (d) Calculate the area of the kite.

13 A triangle has sides of length 5, 12 and 13 cm. Sketch and name the shape formed by the triangle and its image when it is (a) reflected in the 5 cm side, (b) reflected in the 13 cm side, (c) rotated through 180° about the mid-point of the 13 cm side, (d) rotated through 180° about the mid-point of the 12 cm side.

14 (a) Construct full-size the shape shown in Fig. 9.4, which is formed from three semi-circles of diameters 5, 5 and 10 cm. (b) Calculate its area. (c) Calculate its perimeter. (d) Compare your answer to part 'c' with that of the circumference of a circle of radius 5 cm.

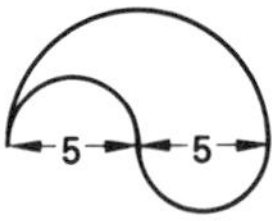

Fig. 9.4

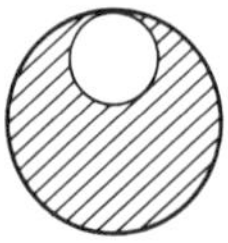

Fig. 9.5

15 The ornament shaded in Fig. 9.5 is formed from two circles of radius 5 and 2 cm respectively.

(a) Draw the figure full size. (b) Calculate the area of gold sheet needed for the ornament. (c) Calculate the cost of platinum wire for the edges at £1.25 per cm.

16 Fig. 9.6 shows a basic tile design and a pattern that can be formed from these tiles.

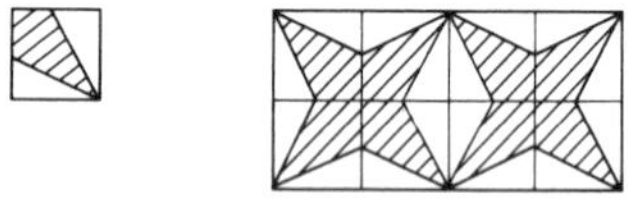

Fig. 9.6

Sketch three other patterns that can be formed from the tiles.

17 Form three different tile patterns from each of the tiles in Fig. 9.7

Fig. 9.7

18 Construct a circle of radius 6 cm. In the circle draw an equilateral triangle with vertices on the circle. Measure its side.

Construct the reflection of the triangle in the centre of the circle. How many axes of symmetry are there in the final figure?

19 Construct a semi-circle of radius 3·5 cm. In this semi-circle construct two triangles on the diameter with angles 30°, 60° and 90°.

Using the centre of the circle as a centre of enlargement construct two similar figures with lengths (a) double and (b) half those of the original.

20 Four goats are tethered in a field, each by a chain 5 m long. G_1 is tethered to a peg, G_2 to a bar of length 12 m, G_3 to a circular groove of radius 10 m and G_4 to a triangular groove with sides 8m.

Construct, to a scale of 1 cm to 2m, the areas over which the goats may graze. Ignore the dimensions of the goats' heads!

21 In each part of Fig. 9.8 m is a main road and F_1, F_2 etc., are farms. Minor roads are to be constructed to link the farms to the main road. In each case only one road junction is permitted.

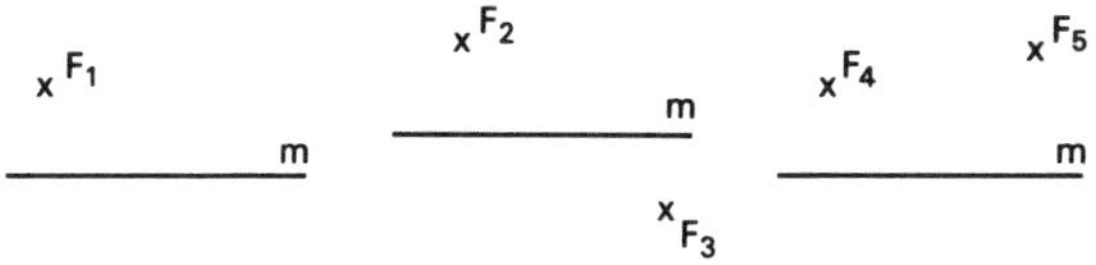

Fig. 9.8

Make three sketches and show, on each, where to place the junction so that the length of minor road is least.

22 In Fig. 9.9, a telephone link is to be made between Headquarters H and Outpost O, crossing the river R at right angles.

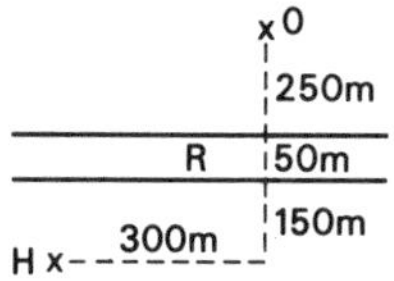

Fig. 9.9

Make a scale drawing of 1 cm to 25 m, and on your drawing indicate the line of the telephone link so as to use least wire.

23 Repeat question 22 for Fig. 9.10, showing how to join well

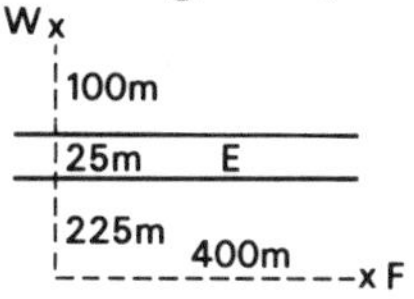

Fig. 9.10

W to farm F, tunnelling through embankment E at right angles and using the least length of water pipe.

Further Geometrical Properties

Exercise 10

For the following questions use Fig. 10.1, *in which O is the centre of the circle and P, Q, R and S are points on its circumference.*

1 If PQ = 8 cm and is 3 cm from the centre O, calculate

(a) the area of triangle OPQ, (b) the length OP, (c) angle POQ, (d) angles PRQ and PSQ.

2 If PQ = 30 cm and OP = 17 cm, calculate (a) the distance of O from PQ, (b) the area of triangle OPQ, (c) angle POQ, (d) angles PRQ and PSQ.

3 If PRQ = 60° and OP = 12 cm, calculate (a) angles PSQ and POQ, (b) length PQ, (c) the area of triangle OPQ.

4 If PQ = 40 cm and its distance from O is 21 cm, calculate (a) length OP, (b) angle POQ, (c) angles PRQ and PSQ.

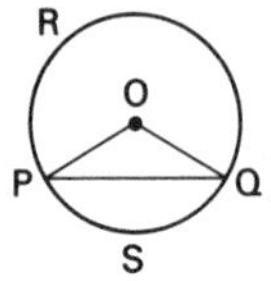

Fig. 10.1

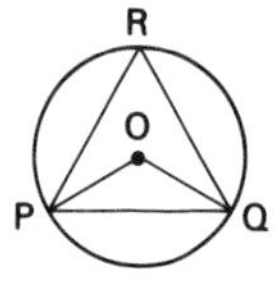

Fig. 10.2

5 Refer to Fig. 10.2 in which O is the centre of the circle and RP = RQ.
(a) If ∠OPQ = 32°, calculate ∠POQ and ∠PRQ. (b) If ∠PRQ = 38°, calculate ∠POQ, ∠OPQ and ∠RQO. (c) If ∠POQ = 132°, calculate ∠OPQ, ∠PRQ and ∠RQO. (d) If ∠RPO = x° and ∠OPQ = $2x$°, calculate x.

6 For these questions use Fig. 10.3 in which O is the centre of the circle and QP = QS.
(a) If ∠QPR = 30°, calculate ∠PRQ, ∠POQ, ∠PQS.
(b) If ∠PRQ = 72°, calculate ∠RPQ, ∠PQS, ∠OPS.
(c) If ∠PSQ = 64°, calculate ∠PQS, ∠PRQ, ∠OPS.
(d) If ∠QPR = x° and ∠PRQ = $2x$°, show that PR divides triangle PQS into two congruent triangles.

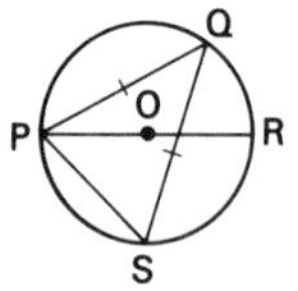

Fig. 10.3

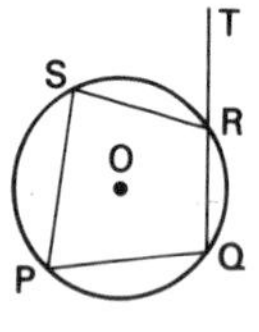

Fig. 10.4

7 Figure 10.4 shows a circle centre O with PQRS a cyclic quadrilateral. Side QR is produced to T.
(a) If ∠SPQ = 75°, calculate ∠SOQ, ∠SRQ, ∠SRT.
(b) If ∠ SRT = 58°, calculate ∠SPQ, ∠SOQ, ∠SRQ.
(c) If ∠SOQ = 140°, calculate ∠SPQ, ∠SRQ, ∠SRT.
(d) Given that ∠SPO = 24° and ∠SRT = 58°, calculate ∠OPQ, ∠PQO, ∠PSQ, ∠OQS.

8 These questions relate to Figure 10.5, in which PA and PB

are tangents to the circle centre O. R and Q are points on the circumference of the circle, as shown.

(a) If ∠ABP = 48°, calculate ∠AOB, ∠AQB, ∠ARB.
(b) If ∠AQB = 72°, calculate ∠AOB, ∠ARB, ∠APB.
(c) If ∠ARB = 112°, calculate ∠APB.
(d) If ∠ABO = 26°, calculate ∠AOB, ∠ARB, ∠APB.
(e) If ∠APB = 50°, calculate ∠OAB.
(f) If ∠AOB = x°, express in terms of x, ∠AQB, ∠ARB, ∠APB.

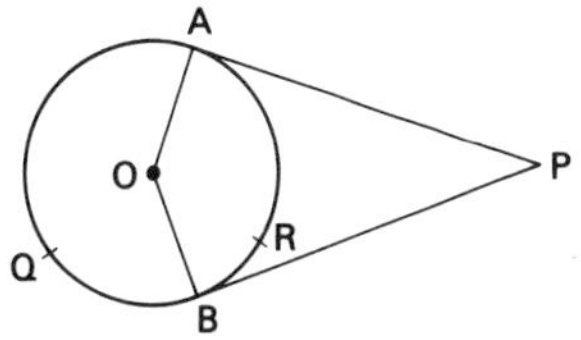

Fig. 10.5

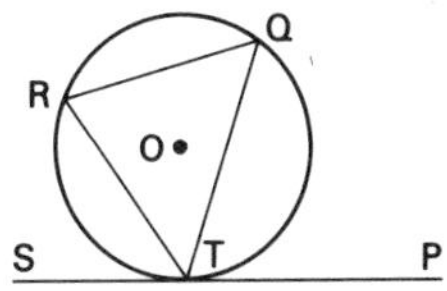

Fig. 10.6

9 Use Figure 10.6, for the following questions. O is the centre of the circle. Calculate:

(a) ∠TRQ, ∠TOQ and ∠OTP, if ∠PTQ = 50°.
(b) ∠TOQ, ∠QTP and ∠OTS, if ∠TRQ = 64°.
(c) ∠TQR, ∠QRT and ∠RTS, if ∠QTP = 60° and PT is parallel to QR.
(d) ∠RTQ, if ∠PTQ = 68° and RT = RQ.

10 For the following questions refer to Fig. 10.7.

(a) Name two pairs of similar triangles in the figure.
(b) Calculate WO if (1) OY = 6 cm, OZ = 4 cm and OX = 12 cm, (2) OY = 5 cm, OZ = OX = 10 cm, (3) OY = 5 cm, OZ = 4 cm and XZ = 15 cm, (4) WY = 17 cm, OZ = 6 cm and OX = 10 cm.

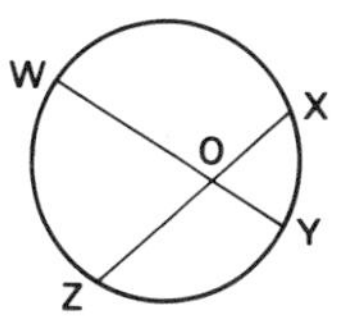

Fig. 10.7

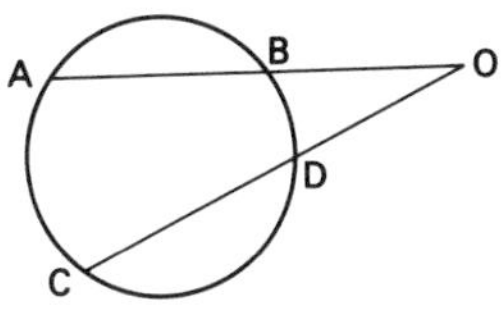

Fig. 10.8

11 Use Fig. 10.8, for these questions.

(a) Explain why triangles ADO and CBO are similar.
(b) Explain why triangles OAC and ODB are similar.
(c) Calculate OB if (1) OA = 24 cm, OD = 8 cm and OC = 18 cm, (2) OA = 20 cm, OD = 6 cm and DC = 8 cm, (3) OA = 16 cm, OC = 12 cm and CD = 9 cm, (4) AB = 6 cm, OD = 5 cm and OC = 8 cm – call OB x and form a quadratic equation.

12 These questions refer to Fig. 10.9.

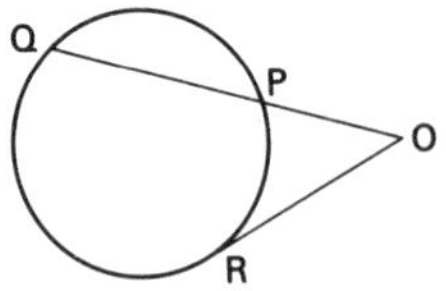

Fig. 10.9

(a) Explain why triangles ORQ and OPR are similar.

(b) Calculate (1) OR, if OP = 8 cm and PQ = 10 cm, (2) OP, if OR = 10 cm and OQ = 15 cm, (3) OR, if OP = 8 cm and OQ = 24 cm, (4) OP, if OR = 8 cm and PQ = 12 cm.

13 A cannonball is almost buried in concrete. The radius of the portion exposed is 6 cm and its highest point is 2 cm above the surface of the concrete. Calculate the radius of the cannonball and the depth of the lowest point below the surface.

14 A humped back bridge is 40 m across and has a radius of curvature of 25m. Calculate the height of its centre above its ends.

15 A hemi-spherical wine glass of radius 6 cm is filled with wine to a depth of 4 cm. Calculate the area of the surface of the wine.

16 A hard ball of radius 15 cm is pushed into soft sand to a depth of 3 cm. Calculate the radius of the circle the ball makes in the sand.

How far should a ball of radius 5 cm be pushed into the sand to give a circle of radius 3 cm?

17 A stream is diverted through a cylindrical drain of radius 52 cm. The width of the top of the water as it flows through is 40 cm. Calculate the two possible depths of the water.

18 Construct squares equal in area to the following rectangles. (a) Sides 9 cm and 5 cm, (b) 8 cm and 3·5 cm, (c) 10 cm and 6 cm, (d) 8·2 cm and 4·3 cm.

In each case measure the side of your square and check by calculation.

19 Construct a triangle of sides 9, 9 and 7 cm. Construct a rectangle equal in area to the triangle and hence a square equal in area to the triangle.

20 Repeat question 19 for triangles of sides (a) 8, 11 and 13 cm, (b) 7, 11 and 12 cm.

21 (a) Construct the quadrilateral PQRS in which PQ = 8 cm, QR = 10 cm, RS = PS = 9 cm and $\angle PQR = 90°$.

(b) Construct a triangle equal in area to the quadrilateral.
(c) Construct a square equal in area to the triangle.

22 Construct a square equal in area to the quadrilateral WXYZ in which WX = 8·4 cm, XY = 7·5 cm, XZ = 8·5 cm, ZW = 6·8 cm and WY = 9 cm.

Three-Dimensional Figures

Exercise 11

Use $\pi = 22/7$ *or* 3·14, *as appropriate.*

1 Calculate the capacity of the following containers.
(a) A fish tank 80 by 40 by 25 cm. (b) An oil drum radius 60 cm, height 140 cm. (c) A spherical tank for liquid oxygen, radius 2·1 m.

2 Find the quantity of wax in ornamental candles shaped as follows. (a) A cube of edge 6 cm. (b) A cuboid 3 by 4 by 6·5 cm. (c) A cylinder radius 4 cm, height 14 cm. (d) A cone radius 4 cm, height 14 cm. (e) A pyramid with base 5 cm square and height 6 cm. (f) A sphere of radius 7 cm.

3 In Fig. 11.1, the lengths are in cm. Sketch the solids formed if the shapes are (a) translated 4 cm at right angles to their plane, (b) rotated through 360° about the line XY.

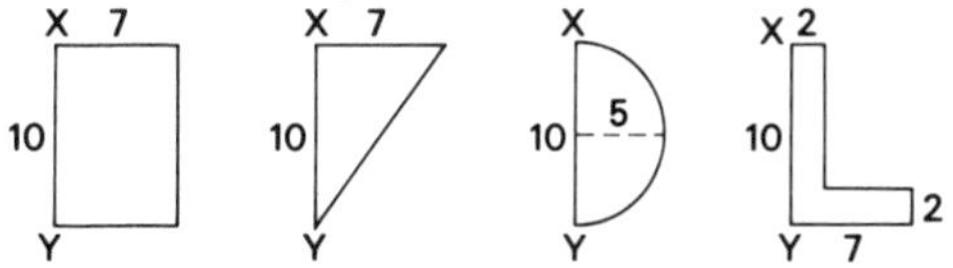

Fig. 11.1

4 Calculate the volumes formed by the transformations in question 3.

5 State the number of centres, axes and planes of symmetry in a prism with (a) a rectangular cross-section, (b) a square cross-section, (c) an equilateral triangle as cross-section.

6 Repeat question 5 for (a) a cylinder, (b) a cone, (c) a sphere, (d) a cube.

7 These questions concern a garden roller of diameter 28 cm and width 60 cm. (a) Calculate how far it rolls forward in one

revolution. (b) Find the area it rolls over in 5 revolutions. (c) Find the weight of the roller if it is solid and weighs 2·1 gm per cm^3. (d) A chalk mark is made from a point X on one edge of the roller to a point Y on the other edge diagonally opposite. Find the shortest length of the chalk line.

8 A rectangle of stiff paper measures 66 by 44 cm. The paper is rolled and its edges taped together to form a cylinder. Calculate (a) the radius of the cylinder if the 66 cm edge is rolled, (b) the volume of the cylinder if the 44 cm edge is rolled.

9 A water trough is in the shape of a half cylinder with diameter 60 cm and length $3\frac{1}{2}$ m. Calculate the capacity of the trough.

Would a trough 3·5 m long with cross section an equilateral triangle of side 60 cm hold more or less water?

10 A practice swimming pool is made from a tank of radius 10·5 m and depth 2 m. Calculate (a) the capacity of the tank, (b) the outside area to be painted if it rests on the ground, (c) the inside area to be coated with water-proofer, (d) the volume of concrete needed to form a base 1m greater in radius than the tank and 14 cm thick.

11 Figure 11.2, shows a cylinder topped by a hemisphere. The lengths are in cm.

(a) Which has the greater volume, the cylinder or the hemisphere? (b) Which has the greater curved surface area? (c) What is the total volume of the solid? (d) What is its total surface area?

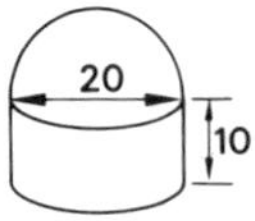

Fig. 11.2

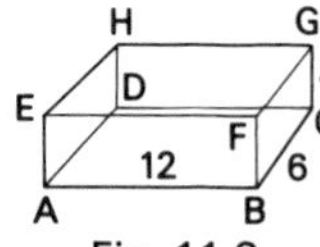

Fig. 11.3

Refer to the cuboid in Fig. 11.3 *for the next three questions. The units are cm.*

12 What is the radius of the largest circles that can be drawn on face ABFE, BCGF, ABCD, ACGE, ABGH?

13 What is the diameter of the largest sphere that would fit into a box of this size? How many such spheres could be fitted? How many spheres of diameter 3 cm would fit?

14 A framework of this shape could be made from two wire rectangles like ABFE and four wire lengths like AD. Six pieces of wire in all. Find a way of constructing the frame with just four pieces of wire. Can it be made with less?

15 What is the maximum fraction of space occupied when (a) a cone is placed in a cylinder, (b) a cylinder is placed in a square prism, (c) a sphere is placed in a cube, (d) a sphere is placed in a cylinder with height equal to the diameter of the sphere? Leave π in your answers where appropriate.

16 Repeat question 15 for (a) a triangular prism placed in a rectangular prism, (b) a pyramid placed in a prism, (c) a cube placed in a cuboid of sides 8, 9 and 10 cm.

17 Copy and complete the following table of scales for similar figures.

Scale for lengths	6:1	3:1	1:12	5:3				
Scale for areas	36:1				1:64		49:4	
Scale for volumes	216:1					1:125		1:1 000

18 A model vintage car is made to 1/24th scale for lengths. What is the scale for (a) the diameter of the steering wheel, (b) the area of the windscreen, (c) the capacity of the boot, (d) the circumference of the tyres, (e) the volume of the cylinders?

19 Three similar saucepans are 10, 12 and 18 cm high. State, in simplest form, the ratio of (a) the lengths of their handles, (b) the area of their bases, (c) their capacities.

20 A drum of salt 15 cm high is said to contain 300 pinches of salt. How many pinches are there in a similar drum 20 cm high?

21 A bottle of rum, 12 cm high, gives 18 tots. How many tots are their in a bottle, similarly shaped, but only 8 cm high?

22 A model village is built to a linear scale of 1 in 20. What is the scale for (a) the area of the village green, (b) the height of the church spire, (c) the quantity of water in the duck pond, (d) the amount of glass in the post office window, (e) the quantity of hay in a hay stack?

23 A giant tin of beans is made for an advertising campaign. The normal tin, 12 cm high, holds 1 272 beans. How many beans does the giant tin, 1 m high, hold?

The outsize tin has a label 3 m^2. What is the area of the normal label?

24 Two similar statuettes have heights in the ratio 5:2. What is the ratio of (a) their areas, (b) their widths, (c) their volumes, (d) their weights, if the statuettes are solid?

The smaller costs £20. What is the price of the larger if it depends 50% on heights, 40% on areas and 10% on volumes?

25 A jet liner flies so successfully that it is decided to build a similar aircraft with all the lengths doubled. Suggest reasons why this double jet would not be successful.

26 Draw nets of (a) a cube, (b) a cuboid, (c) a triangular prism, (d) a cylinder – closed, (e) a pyramid with a square base, (f) a regular tetrahedron and (g) a cone.

27 Say which of the nets in Fig. 11.4, can be folded to form an open cube.

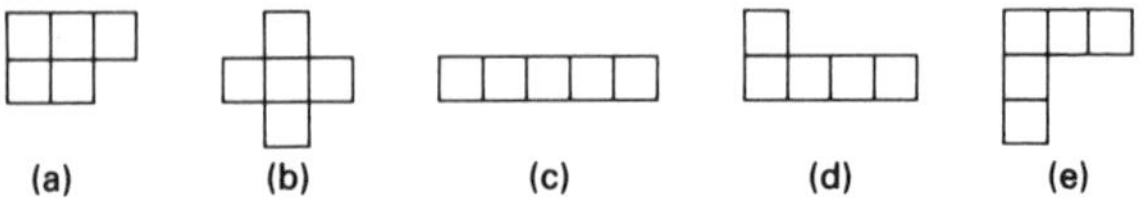

Fig. 11.4

Draw two other 5-square nets that form an open cube.

28 Say which of the nets in Fig. 11.5, can be folded to form a closed pyramid.

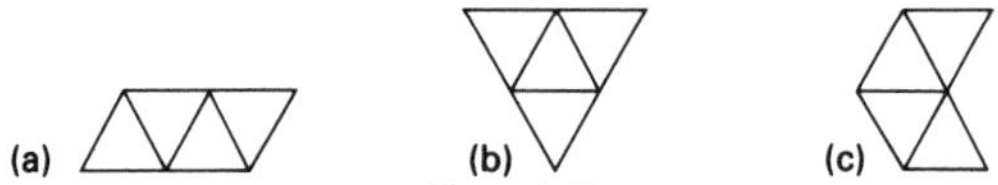

Fig. 11.5

29 Plastic footballs, 24 cm in diameter, are packed in two types of boxes. One is a cube, the other a triangular prism. (a) Sketch nets of the two types of box. (b) Decide which type of box uses less cardboard. Supply evidence to support your answer.

Use Fig. 11.6 *for questions 30–35. The lengths are in cm and the height of X above AD is* $8\frac{1}{2}$ *cm. XY is centrally placed above the plane EFGH.*

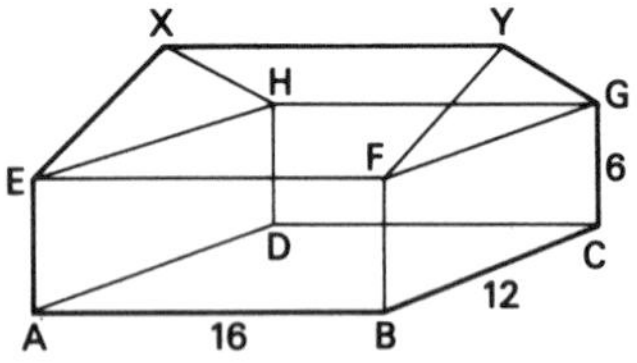

Fig. 11.6

30 Classify the following triangles as acute, obtuse or right angled; isosceles or scalene. Triangles ABD, EDA, YEH, EGC, XGC, EXY, AYG and XBC.

31 Arrange the following diagonals in order of increasing size. Diagonals AC, AF, AH, AY, AG and AX.

32 Draw, one quarter full size, a net of the figure. Use your net to find the shortest distance across the surface from C to E.

33 Draw, half full size, a plan and two elevations of the solid. Measure the angle FYG.

Construct the section through EA and GC.

34 Calculate (a) the volume of the solid, (b) the area of (1) the end BCGYF, (2) the 'roof' section EFYX.

35 Calculate the angles FAB, GAC, FYG and AGB.

36 A wire paper basket is shaped as shown in Fig. 11.7. The measurements are in cm and the basket is part of a pyramid.

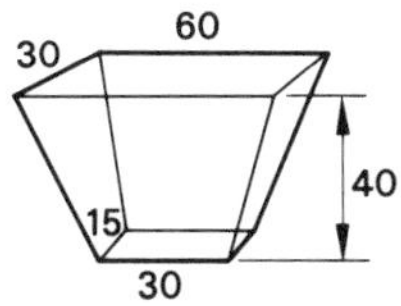

Fig. 11.7

(a) Draw a plan and two elevations of the basket to 1/10th scale.

(b) Measure the angles the sides make with the vertical.

(c) By further drawing – or calculation – find the height of the pyramid of which the basket is part.

(d) Calculate the volume of the basket.

Figure 11.8 *shows the outline of a tent. The lengths are in metres. Use the figure for the following five questions.*

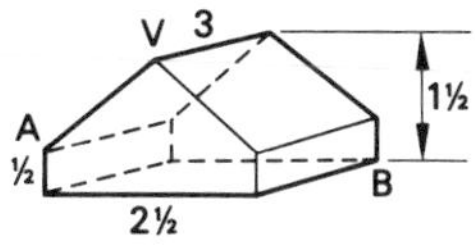

Fig. 11.8

37 Calculate (a) the ground area of the tent, (b) the area of the front end, (c) the volume of air in the tent.

38 Calculate the angle of slope of the roof with the horizontal.

39 Draw a net of the tent to 1/100th scale. From your net decide (a) the minimum number of pieces of canvas needed to make the tent, (b) the shortest distance across the canvas from V to B.

40 Draw a plan and two elevations of the tent to 1/50th scale.

41 Find the radius of the largest beach ball that could be rolled into the tent without bulging the sides.

The next three questions relate to the pyramid VPQRS in Fig. 11.9. *The pyramid has a rectangular base PQRS,* 12 *cm by* 8 *cm, and a height VO of* 14 *cm. W, X, Y and Z are the mid points of the sloping edges.*

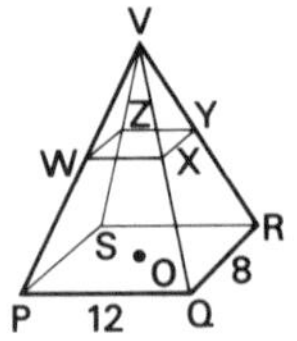

Fig. 11.9

42 Calculate the volumes of the pyramids VPQRS, VWXYZ, VOPQ, VOQR, WPQS and WOPQ.

What fraction of the volume of the complete pyramid is the figure PQRSWXYZ?

43 (a) Draw, half full size, a plan and two elevations of the pyramid VPQRS.

(b) Measure the angles between the opposite pairs of faces.

(c) Find, by further drawing or calculation, the length PV.

44 (a) Draw, $\frac{1}{4}$ full size, the net of the pyramid VPQRS.

(b) Find the total surface area of the pyramid.

(c) Find the shortest distance across the surface from P to Y.

Use Fig. 11.10 *for the next three questions. It shows a sphere of radius* 5 *cm placed on top of a cylinder of radius* 4 cm *and height* 10 *cm.*

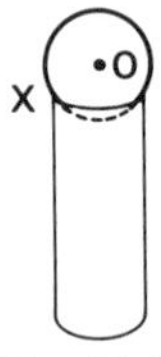

Fig. 11.10

45 (a) Draw, half full size, a plan of the complete figure.
(b) With the same scale draw an elevation – use symmetry and the fact that OX = 5 cm, to draw the sphere.

(c) Measure the height of the centre of the sphere above the base of the cylinder.

46 Calculate the ratio of the volumes of the sphere and the cylinder.

47 The area of the cap of a sphere is given by the formula $A = 2\pi rh$, where 'r' is the radius of the sphere and 'h' the height of the cap. Calculate the area of the sphere that is inside the cylinder.

48 Sketch the solids represented by the plans and elevations drawn in Fig. 11.11.

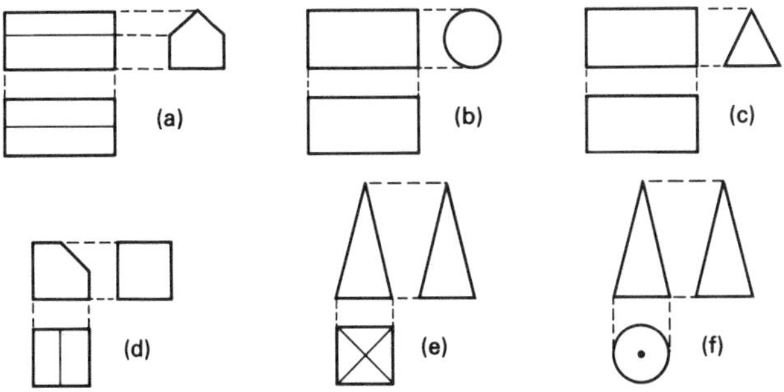

Fig. 11.11

Trigonometrical Ratios (0-360°)

Exercise 12

1 Copy and complete the following table of signs.

	1st	2nd	3rd	4th quadrant.
sin	+			
cos		−		
tan				−

2 (a) Which of the ratios sin, cos or tan are positive in the 1st, 2nd, 3rd and 4th quadrant? (b) Which of the three ratios are negative for 120°, 180°, 220°, 290° and 310°?

3 Find, using tables, sin, cos and tan of 130°, 160°, 170°, 125°, 145°, 165°, 117°, 126°, 157° and 167°.

4 Repeat question 3, for 108° 10′, 123° 20′, 143° 35′, 164° 42′ and 171° 17′.

5 Repeat question 3, for 200°, 250°, 280°, 310°, 345° and 354°.

6 Find the angles $x°$ in the range $0 - 360°$ if $\sin x° = \frac{1}{2}$, $\sin x° = -\frac{1}{2}$, $\cos x° = \frac{1}{\sqrt{2}}$, $\cos x° = -\frac{\sqrt{3}}{2}$, $\tan x° = 1$, $\tan x° = -1$.

7 Repeat question 6, for $\sin x^\circ = 0{\cdot}7$, $\sin x^\circ = -0{\cdot}4$, $\cos x^\circ = 0{\cdot}2$, $\cos x^\circ = -0{\cdot}3$, $\tan x^\circ = 1{\cdot}6$, $\tan x^\circ = -3{\cdot}2$.

8

x°	0°	30°	45°	60°	90°
$\sin x^\circ$				0·87	
$\cos x^\circ$					

(a) Copy and complete the table above, giving answers to 2 decimal places. (b) Extend the table to 120°, 135°, 150° and 180°. (c) Extend both tables to 210°, 225°, 240°, 270°, 300°, 315°, 330° and 360°.

9 Repeat question 8, for $\tan x^\circ$, replacing 90° by 80° and 270° by 260°.

10 Use the results of the previous two questions to plot graphs of $y = \sin x^\circ$, $y = \cos x^\circ$ and $y = \tan x^\circ$ for $0 \leqslant x \leqslant 360$.

11 Sketch graphs of $y = \sin x^\circ$, $y = \cos x^\circ$ and $y = \tan x^\circ$ for x from 0 to 360. Mark on your graphs any axes or centres of symmetry.

12 Sketch the image of $y = \sin x^\circ$ when (a) reflected in the y-axis, the x-axis and the line $x^\circ = 180^\circ$, (b) rotated through 180° about the origin, the point (180°,0) and the point (360°,0).

13 Repeat question 12 for $y = \tan x^\circ$.

14 Sketch the image of the graph of $y = \cos x^\circ$ for $0 \leqslant x \leqslant 360$, when (a) reflected in the x-axis, the y-axis and the line $x = 180$, (b) rotated through 180° about the point (90°,0).

15 Taking $0 \leqslant x \leqslant 360$, sketch graphs of (a) $y = \sin x^\circ$ and $y = -\sin x^\circ$ (b) $y = \cos x^\circ$ and $y = -\cos x^\circ$, (c) $y = \tan x^\circ$ and $y = -\tan x^\circ$. What is the relation between the graphs in each pair?

16 For $0 \leqslant x \leqslant 360$, sketch the graphs (a) $y = \sin x^\circ$, $y = 1 + \sin x^\circ$, $y = 4 + \sin x^\circ$, (b) $y = \cos x^\circ$, $y = \cos x^\circ - 2$, $y = \cos x^\circ - 4$. Comment on the results.

17 With $0 \leqslant x \leqslant 360$, plot graphs of (a) $y = \sin x^\circ$ and $y = \sin (x + 90)^\circ$, (b) $y = \cos x^\circ$ and $y = \cos (x + 90)^\circ$.
Comment on the relations between the graphs.

18 On the same axes plot graphs of (a) $y = \sin x^\circ$, $y = 2 \sin x^\circ$ and $y = 4 \sin x^\circ$, (b) $y = \cos x^\circ$, $y = 3 \cos x^\circ$ and $y = \frac{1}{2} \cos x^\circ$. Use $0 \leqslant x \leqslant 360$ and comment on your results.

19 In the range $0 \leqslant x \leqslant 360$, plot the graphs (a) $y = \sin x^\circ$, $y = \sin 2x^\circ$, and $y = \sin \frac{1}{2}x^\circ$, (b) $y = \cos x^\circ$, $y = \cos 3x^\circ$ and $y = \cos \frac{1}{3}x^\circ$. Make observations on your results.

20 Select the appropriate relation for the graphs in Fig 12.1 from $y = \sin x^\circ$, $y = \cos x^\circ$, $y = \tan x^\circ$, $y = 1 - \sin x^\circ$, $y = 1 + \cos x^\circ$, $y = 1 + \sin x^\circ$ and $y = -\sin x^\circ$.

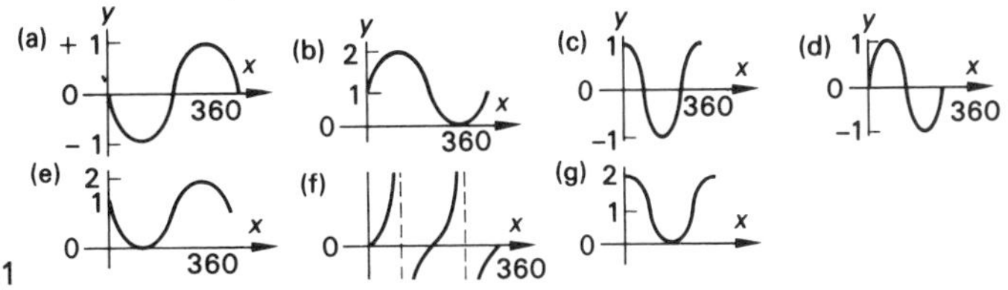

Fig. 12.1

21 Form an equation for each of the graphs in Fig. 12.2.

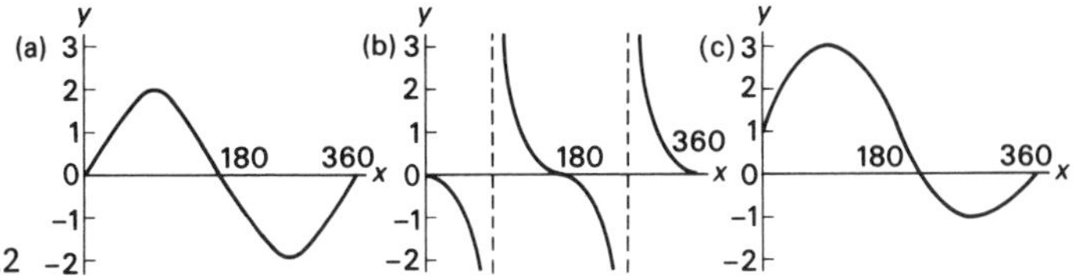

Fig. 12.2

22 Plot graphs of $y = \sin x^\circ$ and $y = \cos x^\circ$ in the range 0 to 360, using one pair of axes. Combine the two graphs to obtain $y = \sin x^\circ + \cos x^\circ$.

23 Use your graphs from question 22 to find (a) $\sin x^\circ + \cos x^\circ$ when $x = 20$, 80, 130 and 320, (b) the maximum and minimum values of $\sin x^\circ + \cos x^\circ$, (c) the solutions of $\sin x^\circ = 0{\cdot}3$, $\cos x^\circ = -0{\cdot}3$ and $\sin x^\circ + \cos x^\circ = 0{\cdot}3$.

24 Plot $y = \sin x^\circ + 2 \cos x^\circ$ for x from 0 to 360.
Use your graph to find (a) any maximum or minimum values of y, (b) solutions of $\sin x^\circ + 2 \cos x^\circ = 0{\cdot}8$.

25 With $0 \leqslant x \leqslant 360$, plot a graph of $y = \sin x^\circ - \cos x^\circ$. Read off (a) any maximum or minimum values of y, (b) the solutions of $\sin x^\circ - \cos x^\circ = 0{\cdot}6$ and $\sin x^\circ - \cos x^\circ = -0{\cdot}6$.

26 Two electrical generators produce power in the pattern $P_1 = 30 \sin 15^\circ t$ and $P_2 = 40 \cos 15^\circ t$ respectively, where t is the time in seconds. Plot a graph of the combined output $P_1 + P_2$ for the first 24 seconds. Read off the greatest positive and greatest negative output and the times at which these occur.

27 Repeat question 26 for $P_1 = 10 \sin 30^\circ t$ and $P_2 = 24 \cos 30^\circ t$, for a time of 12 seconds.

28 On the same axes, sketch the graphs of $y = \tan x^\circ$ and $y = \cos x^\circ$ for $0 \leqslant x \leqslant 360$. How many solutions are there to the equation $\tan x^\circ = \cos x^\circ$, in this range?
Find, by accurate plotting, the smaller of the solutions.

29 By means of a sketch graph, show that there are two solutions to the equation $\sin x^\circ = 2\cos x^\circ$ in the range $0 \leqslant x \leqslant 360$. Find the smaller solution by accurate plotting.

30 (a) Plot $y = \sin 2x^\circ$ for $0 \leqslant x \leqslant 360$. (b) Read off values of y for $x = 20$, 40, 140 and 340. (c) Find the values of x for which $y = 0{\cdot}8$ and $-0{\cdot}8$. (d) By drawing a suitable straight line graph solve $\sin 2x^\circ = \dfrac{x}{200}$.

Trigonometry of the General Triangle

Exercise 13

Areas.

1 Calculate the areas of triangles ABC with dimensions
(1) $a = 9$ cm, $b = 14$ cm and $\angle C = 54^\circ$
(2) $b = 10$ cm, $c = 17$ cm and $\angle A = 77^\circ$
(3) $a = 12$ cm, $c = 11$ cm and $\angle B = 127^\circ$
(4) $a = 7{\cdot}6$ cm, $b = 8{\cdot}3$ cm and $\angle C = 142^\circ$.

2 Calculate and compare the areas of the following pairs of triangles PQR.
(a) $p = 6$ cm, $q = 10$ cm and $\angle R = 50^\circ$; $p = 60$ cm, $q = 100$ cm and $\angle R = 50^\circ$.
(b) $p = 12$ cm, $r = 14$ cm, $\angle Q = 70^\circ$; $p = 12$ cm, $r = 14$ cm, $\angle Q = 110^\circ$.
(c) $q = 20$ cm, $r = 25$ cm, $\angle P = 35^\circ$; $q = 20$ cm, $r = 25$ cm, $\angle P = 145^\circ$.
(d) $p = 7$ cm, $q = 8$ cm, $\angle R = 40^\circ$; $p = 7$ cm, $q = 8$ cm $\angle R = 80^\circ$.

3 Use the formula $A = \frac{1}{2}ab \sin C$, to calculate the areas of the following isosceles triangles. (a) Equal sides 12 cm with an included angle of 50°. (b) Equal sides 15 cm and included angle 70°. (c) Equal sides 18 cm and included angle 130°.

4 Calculate the area of the kite shown in Fig. 13.1. The lengths are in cm.

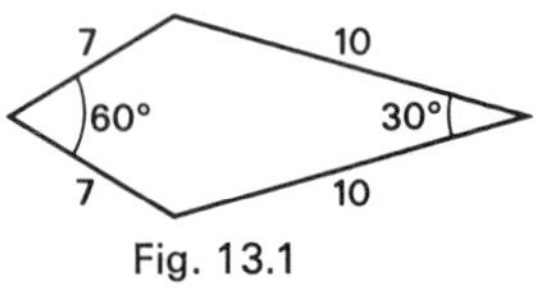

Fig. 13.1

5 (a) Make a sketch of a regular pentagon, drawn in a circle, centre O, radius 10 cm.

(b) Show that the pentagon can be divided into 5 congruent triangles, each with vertex O. What is the value of angle AOB, if AB is one of the sides of the pentagon?

(c) Calculate the area of triangle AOB and hence the area of the pentagon.

6 Repeat question 5 for (a) a regular hexagon, (b) a regular octagon, (c) a regular decagon – 10 sides, drawn in circles radius 10 cm.

7 With a minimum of further calculation, find the areas of the polygons in question 6, if they were drawn in circles of radius 20 cm.

8 Use the formula Area $= \sqrt{s(s-a)(s-b)(s-c)}$, where $s = \frac{1}{2}(a+b+c)$, to calculate the areas of triangles with sides (a) 4, 5 and 7 cm, (b) 6, 8 and 12 cm, (c) 12, 15 and 17 cm, (d) 120, 150 and 170 cm, (e) 1 800, 2 500 and 2 900 m.

9 The three runways on an airport form a triangle of 1 900, 2 100 and 2 300 m. Calculate the area enclosed by the runways – answer in hectares.

10 Calculate the area of the quadrilateral shown in Fig. 13.2. The lengths are in metres.

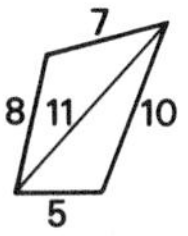

Fig. 13.2

The sine formula.

11 Calculate (1) side a if $\dfrac{a}{\sin 50°} = \dfrac{16}{\sin 40°}$, (2) side b if $\dfrac{b}{\sin 140°} = \dfrac{20}{\sin 70°}$, (3) side c if $\dfrac{c}{\sin 45°} = \dfrac{30}{\sin 105°}$.

12 Calculate the two possible values of angle A if A is in a triangle and (1) $\dfrac{\sin A°}{10} = \dfrac{\sin 70°}{12}$, (2) $\dfrac{\sin A°}{16} = \dfrac{\sin 130°}{20}$.

13 Calculate the larger of the two remaining sides in triangle ABC if (1) $a = 12$ cm, $\angle B = 56°$, $\angle C = 76°$, (2) $b = 15$ cm, $\angle A = 48°$, $\angle C = 36°$, (3) $c = 8$ cm, $\angle A = 105°$, $\angle B = 45°$, (4) $c = 24$ cm, $\angle A = 75°$, $\angle B = 35°$.

14 Use the relation $\dfrac{a}{\sin A} = 2R$, to find the radius R of the circumcircle in a triangle with $a = 16$ cm and $\angle A = 50°$.

15 Repeat question 14 for (1) $a = 20$ cm and $\angle A = 120°$, (2) $b = 15$ cm and $\angle B = 48°$, (3) $c = 300$ m and $\angle C = 110°$.

16 A triangle ABC has side $a = 18$ cm and a circumcircle of radius 10 cm. What are the two possible values of angle A?

17 In triangle LMN, length LM = 18 cm, angle L = 42° and angle M = 32°. Calculate (a) the length LN, (b) the perimeter of the triangle.

18 Searchlight S_1 shines on an aircraft A. At the same time searchlight S_2 picks out the aircraft. The angle of elevation of S_1 is 39° and that of S_2 79°. The distance S_1S_2 is 2 000 m, and S_1S_2A are in the same vertical plane, with S_1 and S_2 on the same side of the aircraft.

Calculate (a) the length AS_1, (b) the height of the aircraft.

19 TV is a television mast – see Fig. 13.3. TS and TR are two wire supports from the top T of the mast to the ground. Angle TSR = 48° and angle TRV = 70°. S, R and V are in a straight line. Length SR = 60 m.

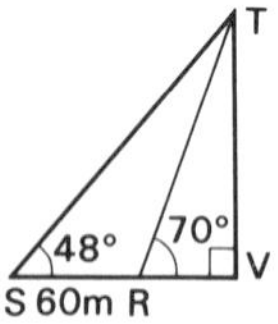

Fig. 13.3

Calculate (a) angles SRT and STR, (b) the length TR, (c) the height TV.

20 A surveyor notes that the angle of elevation of T, the top of a pine tree, is 35°. He walks 50 m directly towards the tree and finds that the angle of elevation is then 55°. Use the method of question 19 to calculate the height of the tree.

21 L is the top of a lighthouse and from a boat the angle of elevation of L is 72°. The boat is moved 40 m directly away from the lighthouse and the angle of elevation is then found to be 42°.

(a) Make a rough sketch of the situation and estimate the height of the lighthouse. (b) Use trigonometry to calculate the height. (c) Use a scale drawing to check the calculations.

22 Observation post P is North East of post O and 800 m away. From O, enemy encampment E bears 020° and from P, E bears 280°.

(a) Make a sketch of the situation, (b) find the angles of the triangle OPE, (c) say which observation post is nearer the enemy, (d) calculate the nearer distance.

The cosine formula.

23 Use the cosine formula to calculate the side a in the triangle ABC when (1) $b = 12$ cm, $c = 8$ cm and $\cos A = \frac{3}{4}$, (2) $b = 7$ cm, $c = 10$ cm and $\cos A = 2/5$, (3) $b = 5$ cm, $c = 9$ cm and $\cos A = \frac{1}{3}$, (4) $b = 6$ cm, $c = 9$ cm and $\cos A = 5/9$.

24 Calculate the third side of the triangle
(a) ABC if $a = 9$ cm, $b = 10$ cm and $\angle C = 50°$,
(b) XYZ if $y = 11$ cm, $z = 15$ cm and $\angle X = 68°$,
(c) PQR if $p = 12$ cm, $r = 18$ cm and $\angle Q = 123°$,
(d) LMN if $l = 20$ cm, $n = 25$ cm and $\angle M = 166°$.

25 Calculate the largest angle in the triangle with sides (a) 4, 5 and 6 cm, (b) 7, 10 and 11 cm, (c) 8, 9 and 14 cm, (d) 200, 300 and 400 m.

26 Find whether the following triangles are acute, obtuse or right-angled. Triangles with sides, (a) 7, 8 and 10 cm, (b) 10, 7 and 4 cm, (c) 15, 20 and 30 cm, (d) 15, 20 and 25 cm (e) 9, 8 and 12 cm.

27 Calculate the perimeter of the triangle with (a) sides 7 and 9 m with an included angle of 40°, (b) sides 7 and 9 m with an included angle of 140°.

28 The diagonals of a parallelogram are 18 and 12 cm long. They intersect at an angle of 60°. Calculate (a) the length of the shorter side of the parallelogram, (b) the length of the longer side, (c) the area of the parallelogram.

29 In Fig. 13.4, the lengths are in metres. Calculate the direction YZ if XY is due North.

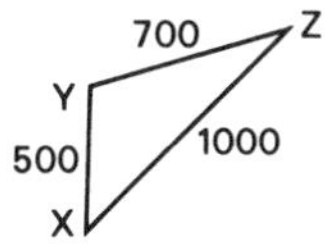

Fig. 13.4

30 From radar station R the bearing and distance of cruiser C is 030°, 14 km. At the same time the bearing and distance of destroyer D is 105°, 18 km. (a) Make a sketch of the situation and estimate the distance DC. (b) Calculate distance DC.

31 At the crack of a gun two runners set off from the same spot. One runs NE at 9 m/s, the other due East at 8 m/s. Calculate the distance between them after 4 seconds.

Find – with minimum further working – the distances after 8 and 10 seconds. Assume constant speeds.

32 'Big Ben' has hands 3·4 and 4·3 m long. What is the distance between the tips of the hands at 1 o'clock?

33 'Little Bill' has hands 0·7 and 1·2 m long. What is the distance between the tips of its hands at 7 o'clock?

34 Which tips of hands are closer together, Bill's at 8 or Ben's at 2 o'clock?

General questions.

35 (a) If sin A = 3/5, express as a fraction cos A and tan A.
(b) Calculate the area of the triangle ABC with $b = 7$ cm, $c = 10$ cm and sin A = 3/5.
(c) Calculate the length of side a.

36 Repeat question 35 for (1) $b = 8$ cm $c = 13$ cm and sin A = 12/13, (2) $b = 17$ cm, $c = 10$ cm and sin A = 15/17.

37 (a) Calculate the three angles of the triangle with sides 5, 6 and 7 metres.
(b) Write down the angles in the triangles with sides (1) 50, 60 and 70 m, (2) 250, 300 and 350 m.
(c) XY = 500 m, YZ = 600 m and ZX = 700 m. If the direction XY is due East, write down the two possible directions of YZ.

38 In Fig. 13.5 the lengths are in cm. Calculate (a) the cosine of angle PRQ, (b) the length of the side TS, (c) the area of the figure PQRST.

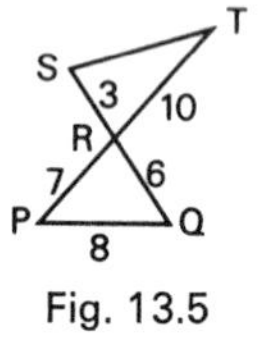

Fig. 13.5

39 Use Fig. 13.6, with lengths in cm and calculate (a) the length PO, (b) the length LM, (c) the area LMNOP.

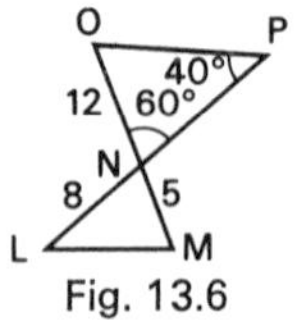

Fig. 13.6

40 The tenth hole on the Tee Side Golf Course is 200 m in length. A golfer hits his drive 140 m, but 8° off the direct line. How far is he then from the hole?

41 The opponent of the golfer in question 40 drives his ball 10° off the direct line. It lands in a bunker, 40 m from the hole. Show that there are four possible positions for the bunker.

Latitude and Longitude

Exercise 14

Take the radius of the earth as 6 360 *km.*

1 Say whether the following are great circles or small circles. The circle 25°N, 52°E, 152°E, 48°S, 148°W, 66°N, the equator, the Tropic of Capricorn.

2 Find the smaller change in longitude between (a) 10°W and 40°W, (b) 10°E and 40°W, (c) 57°E and 75°E, (d) 145°W and 154°W, (e) 145°E and 154°W, (f) 111°E and 111°W.

3 Find the smaller change in latitude between (a) 25°N and 37°N, (b) 52°N and 73°N, (c) 28°S and 28°N, (d) 38°S and the South Pole. (e) 12°S and the North Pole.

4 Calculate the radius of the parallels of latitude 30°N, 45°N, 60°S, 65°S and 77°N.

5 Find the parallels of latitude with radius 3 180, 1 590, 2 000, 3 000 and 720 km.

6 Calculate the lengths of the arc on a circle radius 6 360 km, that contain angles of 60°, 24°, 45°, 72°, 80° and 16°.

7 Calculate the length of the arc that subtends (a) 72° in a circle radius 1 500 km, (b) 48° in a circle radius 1 800 km, (c) 40° with circle radius 2 000 km, (d) 12° with circle radius 2 400 km.

8 Find (a) the change of latitude, (b) the distance from (1) 25°N 40°W to 55°N 40°W, (2) 75°N 20°E to 30°N 20°E, (3) 35°S 40°E to 25°S 40°E, (4) 28°S 58°W to 22°N 58°W.

9 Find (a) the change of longitude, (b) the radius of the parallel of latitude, (c) the distance from (1) 30°N 20°E to 30°N 20°W, (2) 48°S 15°E to 48°S 25°E, (3) 63°N 15°E to 63°N 35°E.

10 Calculate the change of latitude corresponding to arcs of length 2 120 km, 1 060 km, 636 km and 1 908 km.

11 Calculate the change of longitude corresponding to an arc of length, (a) 1 000 km on a circle of radius 3 000 km, (b) 600 km on a circle radius 2 000 km, (c) 750 km on a circle of radius 1 250 km.

12 Find (a) the circumference of the equator, (b) the circumference of the 45th parallel, (c) the parallel of latitude

with a circumference half that of the equator, (d) the parallel of latitude with circumference 1/10th of the equator.

13 Calculate the shortest distance from the parallel 28°N to (a) the equator, (b) the North Pole, (c) the South Pole, (d) the parallel 28°S.

14 'Around the world in eighty days'. What average speed is this following a great circle route?

15 A speed of Mach 2·2 is approximately 2 650 km/h. How long would it take to fly half way round the world on a great circle route at that speed?

What is the saving in flying time with speed of Mach 3?

16 Two aircraft set off from 53°N 28°W. One flies direct to 43°N 28°W and the other direct to 53°N 18°W. Find the distance travelled by each aircraft.

17 Calculate the distance from 80°N 50°E to 80°N 130°W (a) flying along the 80th parallel, (b) flying the great circle route over the North pole.

18 Calculate (a) the radius of the 52nd parallel, (b) the distance from P, 52°N 32°E to Q, 52°N 42°E along the 52nd parallel, (c) the new longitude of an aircraft that flies from P due West for 3 000 km.

19 An aircraft starts from 40°S 10°W. Find its new longitude if it flies (a) 1 200 km due West, (b) 2 400 km due East.

20 A model is made of the earth with a diameter of 2 metres. Calculate for the model (a) the radius of the 18th parallel, (b) the distance from the 18th parallel North to the North Pole, (c) the distance along the parallel from 16°E to 20°W.

21 The moon has a radius of approximately 1 740 km. Calculate (a) the distance from its equator to its poles, (b) the radius of the lunar 50th parallel, (c) the distance along this parallel from 24°W to 48°E.

22 A capsule orbits the moon at a height of 60 km above its surface. Find (a) the circumference of the orbit, assuming it is a great circle, (b) the speed of the capsule, if it completes one orbit every 95 minutes.

Probability

Exercise 15

1 A card is drawn at random from a standard pack of 52 playing cards. What is the probability that it is (a) a red card, (b) a diamond, (c) a picture card, (d) an ace, (e) a black Jack, (f) the King of clubs, (g) not a heart?

2 A six sided die is rolled. What is the probability of (a) a 6, (b) an odd number, (c) a multiple of 3, (d) a prime number, (e) not a 4, (f) not the number which came up the previous time?

3 The shoe sizes of 30 members of a class are distributed as follows:

Size of shoe	3	4	5	6	7	8	9
No of pupils	1	2	5	11	6	3	2.

If a member of the class is chosen at random, what is the probability that their shoe size is (a) 5, (b) 7, (c) 8 or 9, (d) extremely large or extremely small?

4 Nine discs numbered 1 to 9 are placed in a bag and shaken about. One disc is chosen 'blind'. What is the probability that it will be (a) the number 9, (b) an even number, (c) an odd number, (d) a multiple of 3?

A prize of 10p is offered for a 5p stake if a prime numbered disc is chosen. The stake money is not returned. Will the proprietor or the player be more likely to make a profit on this basis? Give reasons for your answer.

5 6000 tickets are sold for a Christmas Raffle. There is one top prize, followed by four major prizes and twenty consolation prizes. What is the probability that a purchaser of 10 tickets will win (a) the top prize, (b) a consolation prize, (c) a prize of some kind, (d) two major prizes?

6 In a boys' school of 600 pupils there are 40 pairs of brothers. What is the chance that a boy chosen at random will have a brother in the same school?

If there are 3 pairs of twins what is the chance that the boy will have a twin brother?

7 A hundred wooden spheres are numbered 1 to 100, placed in a bag and shaken. One sphere is selected. Say whether the first or the second of the following events is more likely, or whether the two events are equally likely.

(a) An odd number or an even number sphere.

(b) A multiple of 10 or a multiple of 5. (c) A two-figure number or a one-figure number. (d) A number ending in 4 or a number ending in 6. (e) A perfect square or a perfect cube. (f) A number with figures adding up to 6 or adding up to 7. (g) A number with two figures both the same or a number with one figure only.

8 The 26 letters of the alphabet are printed each on a separate card. The cards are laid face down on a table and mixed. One card is now turned face upward. What is the chance that it is (a) the letter Z, (b) a vowel, (c) a letter in the word RANDOM, (d) a letter in the word ALPHABET?

9 The letters of the word IGLOO are written one to a card. One card is selected as in question 8. What is the probability the letter is (a) a vowel, (b) an O, (c) the letter L, (d) not the letter G?

10 A pair of cards is selected from the five cards of question 9. What is the probability that (a) both are vowels, (b) both are O's, (c) neither is a vowel, (d) not both of them are vowels?

11 A match is thrown on to a table top. What is the probability that its head points in a direction between (a) North and East, (b) South and South West, (c) South and North East? Take the smaller angle in each case.

12 Two matches are thrown onto a table. What is the probability that (a) both heads point between North and East, (b) one at least points between North and East, (c) one only points between North and East?

13 The four Kings from a pack of playing cards are laid face down on a table and mixed. One King is now selected. What is the chance that it is (a) a red King, (b) the King of Spades?

14 A pair of Kings from the cards in question 13 is now selected. What is the chance that (a) both are red, (b) one is the King of Clubs and the other of Hearts?

15 Two cards are selected at random from a pack of 52 playing cards. What is the probability that (a) the first card only is an ace, (b) the second card only is an ace, (c) both cards are aces, (d) at least one of the cards is an ace?

16 A six sided die is rolled and a penny is tossed. What is the probability of (a) a six on the die, (b) a head on the penny, (c) a six on the die and a head on the penny, (d) a six on the die or a head on the penny (or both)?

17 Two six-sided dice are rolled. Make a table of the 36 possible pairs of scores.

What is the probability of (a) a total of 12, (b) a total of 6, (c) a double, (d) two odd numbers?

What is the most likely total? What is the most likely difference between the two scores?

18 A pointer (see Fig. 15.1) spins around and comes to rest over one of the numbers from 1 to 6 marked on a regular hexagon. Find the probability of (a) a score of 6, (b) an even-numbered score, (c) a score of less than 3, (d) two successive odd-numbered scores, (e) a total of 10 in two spins, (f) a total of at least 10 in two spins.

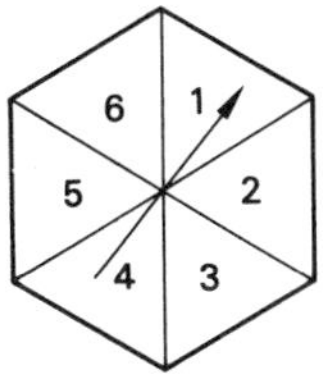

Fig. 15.1

19 A pointer, similar to that in question 18, spins over a regular pentagon with spaces numbered 1 to 5.

(a) Which is more likely: an odd numbered score or an even? (b) In two spins is an even or an odd-numbered total more likely? (c) What is the probability of a 5, of two successive 5's, of three successive 5's?

A prize of £5 is offered for a 1p stake if 5 successive 5's are spun. Discuss whether this is a fair offer.

20 A disc of diameter 1 cm is 'shoved' across a board ruled with lines 3 cm apart (Fig. 15.2).

What is the probability that the disc will end up between a pair of lines?

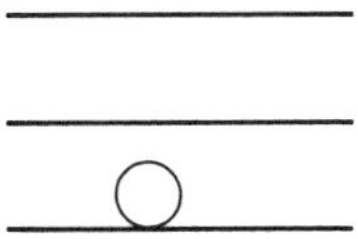

Fig. 15.2

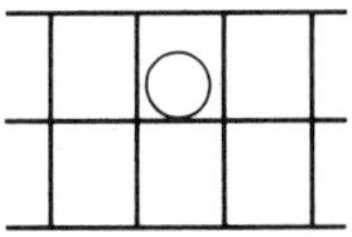

Fig. 15.3

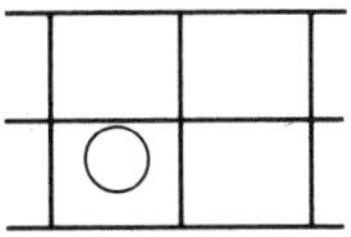

Fig. 15.4

21 Repeat question 20 for a board ruled in squares of side 3 cm (Fig 15.3).

22 Repeat question 20 for a board ruled in rectangles 3 cm by 5 cm (Fig. 15.4).

23 What are the new probabilities if a disc of diameter 2 cm is 'shoved' across the boards in questions 20, 21 and 22?

24 Careful observation of a fairground game reveals that the probabilities associated with the game are:

Event:	Lose 1p stake	2p returned	5p returned	1p stake returned
Probability:	0·4	0·2	0·1	0·3

What is the sum of these four probabilities?

What is the chance of (a) gaining 8p in 2 attempts, (b) losing 2p in 2 attempts, (c) gaining 5p in 2 attempts, (d) neither losing nor winning in 2 attempts?

Is this game likely to prove profitable to its owner? Give reasons to support your answer.

Statistics

Exercise 16

1 A random sample of 50 families is required. Say which of the following samples might be biased and explain why. (a) Families chosen from one housing estate. (b) Families chosen at random from the telephone directory. (c) Families chosen from cars entering a car park. (d) Families chosen from a list of colour television licence holders.

2 Suggest how to make a random sample of (a) trees in a wood, (b) weeds in a playing field, (c) light bulbs from a day's production, (d) spectators at a 1st Division match, (e) hay from a haystack, (f) water from a canal.

3 Draw pie charts to illustrate the following:

(1) The viewing time of a family is distributed as BBC 1:11 h BBC 2:5 h ITV:8 h.

(2) The average expenditure by a representative on travel. Bus £1.60, Train £4.80, Car £4.20, Taxi £1.40.

(3) The areas, in hectares, devoted to four types of tree. Spruce 120 000, Pine 90 000, Beech 22 000, Oak 8 000.

4 The table below shows the estimated number of visitors to a seaside resort in thousands.

1965	'66	'67	'68	'69	'70	'71
210	225	245	220	216	190	185.

Illustrate the data by means of (a) bar chart, (b) a frequency polygon.

5 Use the data of question 4 to calculate the mean number of annual visitors to the resort. In which years has the number of visitors fallen below the mean?

6 Arrange the following sets of data in ascending order. State the range in each case.

(a) Number of eggs laid by seven hens in a week: 5, 8, 2, 7, 6, 9, 5. (b) Scores at darts: 37, 46, 21, 72, 13. (c) Number of days holiday per year: 18, 21, 17, 27, 43, 32, 12, 22. (d) Number of peas in a pod: 6, 8, 6, 5, 7, 4, 3, 5. (e) A day's production of Chocolate Bars: 140 000, 120 000, 130 000, 110 000, 150 000.

7 Find the median values of the sets of data in question 6.

8 Use the data of question 6 and calculate the mean of each set.

9 State the modal class and estimate the median measurement for the following distributions:

(a)

x	10	11	12	13	14
f	2	5	8	7	3

(b)

x	5	7	9	11	13	15
f	4	6	7	8	12	3.

(c)

x	0	4	8	12	16	20	24	28
f	3	8	11	28	20	15	13	2

(d)

x	2·0	2·1	2·2	2·3	2·4	2·5
f	3	6	11	10	8	2

10 Illustrate the data in question 9 by (1) a bar chart for part a, (2) a pie chart for part b, (3) a histogram for part c, (4) a frequency polygon for part d.

11 Calculate the mean measurements for the sets of data in question 9. Use an assumed mean where this is convenient.

12 A farmer has grown a field of carrots. He takes a sample of 36 of these carrots chosen from different parts of the field. He measures the lengths of these carrots and records them as follows:

15	18	14	17	16	15	14	15	15	15	16	19
14	12	13	15	15	14	16	14	14	16	18	17
13	15	16	13	14	16	15	14	15	12	17	13 cm.

(a) Make a tally of the carrots at 1 cm intervals. (b) What is the length of (1) the longest, (2) the shortest carrot? (c) What is the modal length? (d) What is the median length?

13 Illustrate the data in question 12 by means of (a) a bar chart, (b) a pie chart.

14 Calculate the mean length of the carrots in the sample in question 12.

15 Another carrot is chosen at random from the field in question 12. Assuming the lengths are distributed as in the sample, what is the probability of choosing a length of (a) 19 cm, (b) 12 cm, (c) 13 cm or less, (d) 16 cm or more?

(e) What is the probability of choosing two carrots both of length 19 cm?

16 40 cadets were trained to dismantle a gun. On a trial their times for the operation were:

106	108	95	102	108	103	108	120	112	117
92	105	109	127	100	109	113	96	115	110
105	101	114	109	98	107	103	122	119	110
118	111	101	106	126	114	99	104	113	124 seconds.

(a) Make a tally of these times using classes 90–94, 95–99. . . . . 125–129 seconds. (b) What is the modal class? (c) What is the median time? (d) What percentage of the cadets took 2 or more minutes to dismantle the gun?

17 Illustrate the data in question 16 by means of (a) a histogram (b) a frequency polygon. Which method is more suitable for comparing the present trial with one made two weeks earlier?

18 Calculate the mean time for dismantling the guns in question 16.

19 The milometer readings of 100 cars in a car park were distributed as follows:

0–5 999	6 000–11 999	12 000–17 999	18 000–23 999
8	18	28	32
24 000–29 999	30 000–35 999	36 000–41 999	
20	16	11	
42 000–47 999	48 000–53 999	54 000–59 999	
9	5	2	
60 000–65 999			
1			

(a) Illustrate the data by a histogram. (b) State the modal class. (c) Obtain an estimate of the mode.

20 Calculate the mean car mileage for the data in question 19.

21 On the evidence of the data in question 19, what is the chance that the next car to enter will have a mileage of (a) less than 6 000, (b) less than 24 000, (c) more than 47 999?

22 A sample of 150 books was chosen from various shelves in a public library. The number of times they had been borrowed in the past year was distributed as follows:

No of times borrowed	0–4	5–9	10–14	15–19	20–24	25–29	30–34	35–39
Frequency	8	16	29	37	28	17	12	3

(a) Illustrate the data by means of a bar chart. (b) State the modal class and the range. (c) Calculate the percentage of books borrowed less than 10 times in the year.

23 For the data in question 22 (a) list the class marks, (b) estimate the mean, (c) use your estimated mean to help calculate the actual mean.

24 Compile a cumulative frequency table for the data in question 23 and plot an ogive. From your curve obtain an estimate of (a) the median, (b) the semi-interquartile range.

25 A pair of dice are rolled 145 times and the frequencies of the eleven possible total scores were recorded as

Total Score:	2	3	4	5	6	7	8	9	10	11	12
Frequency:	5	9	12	15	21	26	19	16	10	8	4

Illustrate the data by means of (a) a histogram, (b) a frequency polygon.

26 For the data in question 25, state (a) the modal score, (b) the median score.
Calculate the mean score.

27 Compile a cumulative frequency table for the scores in question 25 and plot an ogive. Use your curve to estimate the semi-interquartile range.

28 A survey was made of 240 houses on an estate. The number of occupants in each house ranged from 0 to 10 and the frequencies with which these occurred is shown below:

No of occupants:	0	1	2	3	4	5	6	7	8	9	10
No of houses:	4	18	24	36	56	50	28	10	7	5	2.

Illustrate the data by (a) a pie chart, (b) a histogram.

29 For the data in question 28, state (a) the mode, (b) the median.
Calculate the mean number of occupants.

30 If one of the houses in question 28 is chosen at random what is the probability that it will have (a) 3 occupants, (b) 7 occupants, (c) 7 or more occupants?
What is the probability that it is not empty?

31 Make a cumulative frequency curve from the data in question 28 and use your curve to estimate (a) the median, (b) the semi-interquartile range.

32 The average age of the 17 girls in a class is 15 years 4 months. The average age of the 13 boys in the class is 15 years 6 months. Calculate the average age of the 30 members of the class.

33 The mean length of a sample of 32 vegetable marrows was 72 cm. The mean length of a second sample of 48 marrows was 68 cm. Calculate the mean length of the 80 marrows.

34 *Size of Sample*: 800 700 500
Mean Income: £1 020 £1 300 £1 400

The table above shows the mean incomes of three samples of the population of a country town. Calculate the mean income of the 2 000 people.

Matrices

Exercise 17

1 $\mathbf{M} = \begin{pmatrix} 1 & 1 \\ 2 & 2 \\ 3 & 3 \end{pmatrix}$ $\mathbf{N} = \begin{pmatrix} 1 & 3 & 5 \\ 2 & 4 & 6 \end{pmatrix}$

(a) Write down the orders of matrices **M** and **N**. (b) Write down the order of the product **MN**. (c) Calculate **MN**. (d) Is it possible to calculate **NM**? (e) If so does $\mathbf{MN} = \mathbf{NM}$? (f) Are all of the following impossible? $\mathbf{M} + \mathbf{N}$, $\mathbf{M} - \mathbf{N}$, $2\mathbf{N}$, $\mathbf{M}^2$.

2 $\mathbf{W} = \begin{pmatrix} 1 & 2 & 3 \\ 4 & 5 & 6 \end{pmatrix}$ $\mathbf{X} = \begin{pmatrix} 1 \\ 4 \\ 9 \end{pmatrix}$ $\mathbf{Y} = \begin{pmatrix} 2 & 6 \\ 3 & 6 \end{pmatrix}$ $\mathbf{Z} = (3 \quad 4)$.

(a) Write down the order of each matrix. (b) State the orders of **WX**, **ZY** and **YW**. (c) Calculate **WX**, **ZY** and **YW**. (d) Say whether the following products are possible: **XY**, **YZ**, **WZ** and **ZW**. (e) Is it possible to add any pair of these matrices? (f) Which one of the following is possible? $\mathbf{W}^2$, $\mathbf{X}^2$, $\mathbf{Y}^2$, $\mathbf{Z}^2$.

3 What can be said about matrix **K** if it is possible to calculate $\mathbf{K}^2$? Give an example of such a matrix.

4 $\mathbf{K} = \begin{pmatrix} 3 & 4 \\ 4 & 3 \end{pmatrix}$ $\mathbf{L} = \begin{pmatrix} 1 & 2 & 1 \\ 2 & 1 & 2 \end{pmatrix}$ $\mathbf{M} = \begin{pmatrix} 1 & 2 \\ 3 & 3 \\ 2 & 1 \end{pmatrix}$

Where possible, calculate **KL**, **LK**, **KM**, **MK**, **ML**, **LM**.

5 $\mathbf{K} = \begin{pmatrix} 3 & 4 & 5 \\ 2 & 1 & 6 \end{pmatrix}$ $\mathbf{L} = \begin{pmatrix} 4 & 3 \\ 2 & -1 \end{pmatrix}$ $\mathbf{M} = \begin{pmatrix} 1 & 3 \\ 2 & 4 \\ -6 & 2 \end{pmatrix}$

$\mathbf{N} = \begin{pmatrix} 6 & 0 & 2 \\ 4 & -3 & 1 \end{pmatrix}$

Find, where possible, $\mathbf{K} + \mathbf{M}$, $\mathbf{K} - \mathbf{N}$, **KL**, **LK**, **KM**, **LM**, $4\mathbf{M}$, $2\mathbf{L}$, $3\mathbf{K} - 2\mathbf{N}$, $\mathbf{L}^2$, $\mathbf{K}^2$, $\mathbf{M}^2$, $(\mathbf{KM})^2$.

6 **S** is a 2×2 matrix, **T** is a 3×2 matrix and **R** is another 3×2 matrix. Say if it is possible to calculate $\mathbf{R} + \mathbf{T}$, $\mathbf{T} + \mathbf{R}$, **ST**, **TR**, **RT**, **4T**, $\mathbf{R}^2$, $\mathbf{S}^2$.

7 $\mathbf{M} = \begin{pmatrix} 4 & 3 & 2 \\ 3 & 1 & 2 \\ 1 & 5 & 3 \\ 6 & 1 & 4 \end{pmatrix}$ Find (a) $(1 \quad 1 \quad 1 \quad 1)\mathbf{M}$,
(b) $\mathbf{M}\begin{pmatrix} 1 \\ 1 \\ 1 \end{pmatrix}$.

What is the effect of these operations?

Calculate $[(1\ 1\ 1\ 1)\mathbf{M}]\begin{pmatrix} 1 \\ 1 \\ 1 \end{pmatrix}$ and $(1\ 1\ 1\ 1)\left[\mathbf{M}\begin{pmatrix} 1 \\ 1 \\ 1 \end{pmatrix}\right]$

Comment on your answers.

8 Use the matrix $\begin{pmatrix} 3 & 1 \\ 5 & 2 \end{pmatrix}$ to code the words

BOMB, FIRE, THUD.
What matrix would act as a decoder?

9 (a) Use the matrix from question 8 to code the words DICE, MAXI COAT, MINI CAR.
(b) $\begin{pmatrix} 28 & 50 \\ 52 & 85 \end{pmatrix}$ is a word that has been coded by the above matrix. Find the word.

10 Decode the following words using $\begin{pmatrix} 4 & -1 \\ -3 & 1 \end{pmatrix}$ as a decoder.
(a) $\begin{pmatrix} 27 & 29 \\ 100 & 95 \end{pmatrix}$, (b) $\begin{pmatrix} 16 & 14 \\ 51 & 47 \end{pmatrix}$, (c) $\begin{pmatrix} 20 & 21 \\ 72 & 79 \end{pmatrix}$.

11 Write down the inverses of the following matrices:
$\begin{pmatrix} 5 & 2 \\ 7 & 3 \end{pmatrix}$, $\begin{pmatrix} 8 & 3 \\ 5 & 2 \end{pmatrix}$, $\begin{pmatrix} 7 & 4 \\ 5 & 3 \end{pmatrix}$, $\begin{pmatrix} 3 & 5 \\ 1 & 2 \end{pmatrix}$, $\begin{pmatrix} 3 & 2 \\ 4 & 3 \end{pmatrix}$,
$\begin{pmatrix} -6 & 5 \\ -5 & 4 \end{pmatrix}$, $\begin{pmatrix} -3 & 5 \\ -2 & 3 \end{pmatrix}$, $\begin{pmatrix} 4 & 3 \\ -7 & -5 \end{pmatrix}$.

12 Where possible, write down the inverses of the following:
$\begin{pmatrix} 6 & 9 \\ 5 & 8 \end{pmatrix}$, $\begin{pmatrix} 3 & 4 \\ 4 & 6 \end{pmatrix}$, $\begin{pmatrix} 8 & -3 \\ -4 & 2 \end{pmatrix}$, $\begin{pmatrix} 6 & 3 \\ 8 & 4 \end{pmatrix}$, $\begin{pmatrix} 3 & 8 \\ 2 & 5 \end{pmatrix}$,
$\begin{pmatrix} 4 & 2 \\ 5 & 6 \end{pmatrix}$, $\begin{pmatrix} -8 & 10 \\ 5 & -7 \end{pmatrix}$, $\begin{pmatrix} -5 & -3 \\ 10 & 6 \end{pmatrix}$.

13 $\mathbf{S} = \begin{pmatrix} 2 & 5 \\ 1 & 3 \end{pmatrix}$ $\mathbf{T} = \begin{pmatrix} 5 & 4 \\ 2 & 3 \end{pmatrix}$, find the inverses of:

S, **T**, **2S**, **3T**, $\mathbf{S} + \mathbf{T}$, $\mathbf{S} - \mathbf{T}$, **ST**.

14 $\mathbf{P} = \begin{pmatrix} 2 & 1 \\ 0 & 3 \end{pmatrix}$ $\mathbf{Q} = \begin{pmatrix} 3 & 2 \\ 4 & 1 \end{pmatrix}$ $\mathbf{R} = \begin{pmatrix} 1 & 5 \\ 2 & 4 \end{pmatrix}$.

Find (a) $(\mathbf{P} + \mathbf{Q}) + \mathbf{R}$ and $\mathbf{P} + (\mathbf{Q} + \mathbf{R})$
(b) $(\mathbf{P} - \mathbf{Q}) - \mathbf{R}$ and $\mathbf{P} - (\mathbf{Q} - \mathbf{R})$
(c) $(\mathbf{PQ})\mathbf{R}$ and $\mathbf{P}(\mathbf{QR})$.

15 Does the evidence of question 14 justify the following statements? (a) Subtraction of matrices is not associative. (b) Addition of matrices is always associative. (c) The evidence is consistent with the multiplication of matrices being associative.

16 Using $\mathbf{S} = \begin{pmatrix} 2 & 5 \\ 3 & 6 \end{pmatrix}$ and $\mathbf{T} = \begin{pmatrix} 1 & 4 \\ 2 & 5 \end{pmatrix}$,

find $\mathbf{S} + \mathbf{T}$, $\mathbf{S} - \mathbf{T}$, $\mathbf{ST}$, $\mathbf{TS}$, $\mathbf{S}^2$ and $\mathbf{T}^2$.
(a) Does $(\mathbf{S} + \mathbf{T})(\mathbf{S} - \mathbf{T}) = \mathbf{S}^2 - \mathbf{T}^2$?
(b) Does $(\mathbf{S} + \mathbf{T})(\mathbf{S} - \mathbf{T}) = \mathbf{S}^2 + \mathbf{TS} - \mathbf{ST} - \mathbf{T}^2$?
(c) Does $\mathbf{ST} - \mathbf{T}^2 = \mathbf{T}(\mathbf{S} - \mathbf{T})$? (d) Can you think of a way of factorising $\mathbf{ST} - \mathbf{T}^2$?

17 Taking $\mathbf{L} = \begin{pmatrix} 4 & 7 \\ 3 & 2 \end{pmatrix}$ and $\mathbf{M} = \begin{pmatrix} 3 & 1 \\ 4 & 1 \end{pmatrix}$

find $\mathbf{L} + \mathbf{M}$, $\mathbf{L}^2$, $\mathbf{M}^2$, $\mathbf{LM}$ and $\mathbf{ML}$.
(a) Show that $(\mathbf{L} + \mathbf{M})^2$ does not equal $\mathbf{L}^2 + 2\mathbf{LM} + \mathbf{M}^2$.
(b) Show that $(\mathbf{L} + \mathbf{M})^2 = \mathbf{L}^2 + \mathbf{LM} + \mathbf{ML} + \mathbf{M}^2$.

18 Simplify (a) $\begin{pmatrix} 3 & 0 \\ 4 & 0 \end{pmatrix}\begin{pmatrix} 0 & 0 \\ 5 & 2 \end{pmatrix}$ (b) $\begin{pmatrix} 4 & 2 \\ 8 & 4 \end{pmatrix}\begin{pmatrix} -4 & 2 \\ 8 & -4 \end{pmatrix}$

(c) $\begin{pmatrix} 3 & 5 \\ 6 & 10 \end{pmatrix}\begin{pmatrix} -10 & -5 \\ 6 & 3 \end{pmatrix}$.

Make up two more matrix products that equal $\begin{pmatrix} 0 & 0 \\ 0 & 0 \end{pmatrix}$.

19 From the evidence of question 18, does $\mathbf{AB} = \mathbf{O}$ imply that $\mathbf{A} = \mathbf{O}$ or $\mathbf{B} = \mathbf{O}$?
Does $\mathbf{A} = \mathbf{O}$ and $\mathbf{B} = \mathbf{O}$ imply $\mathbf{AB} = \mathbf{O}$?
What happens if $\mathbf{A} = \mathbf{O}$ and $\mathbf{B}$ is not equal to $\mathbf{O}$?

20 If $\mathbf{M} = \begin{pmatrix} 3 & 2 \\ 5 & 4 \end{pmatrix}$ then $\mathbf{M}^t = \begin{pmatrix} 3 & 5 \\ 2 & 4 \end{pmatrix}$ is called 'the transpose of $\mathbf{M}$'. It is formed by transposing rows into columns.

Form the transpose of $\begin{pmatrix} 6 & 3 \\ 8 & 1 \end{pmatrix}$, $\begin{pmatrix} 4 & 2 \\ -3 & 5 \end{pmatrix}$, $\begin{pmatrix} 7 & 8 \\ 2 & 1 \end{pmatrix}$,
$\begin{pmatrix} 8 & -3 \\ 2 & 5 \end{pmatrix}$, $\begin{pmatrix} 4 & -7 \\ -2 & 3 \end{pmatrix}$.

21 $\mathbf{P} = \begin{pmatrix} 6 & 2 \\ 1 & 3 \end{pmatrix}$ $\mathbf{Q} = \begin{pmatrix} 4 & 1 \\ 7 & 2 \end{pmatrix}$.

(a) Find $\mathbf{P}^t$ and $\mathbf{Q}^t$. (b) Calculate $\mathbf{PQ}$ and $(\mathbf{PQ})^t$.
(c) Calculate $\mathbf{P}^t\mathbf{Q}^t$. (d) Does $(\mathbf{PQ})^t = \mathbf{P}^t\mathbf{Q}^t$?
(e) Does $(\mathbf{PQ})^t = \mathbf{Q}^t\mathbf{P}^t$?

22 Find the matrices $\mathbf{Q}$ if (a) $\mathbf{P} = \begin{pmatrix} 4 & 1 \\ 2 & 3 \end{pmatrix}$ and $\mathbf{P} + 2\mathbf{Q} = \mathbf{P}^2$.
(b) $\mathbf{P} = \begin{pmatrix} 3 & -2 \\ -1 & 4 \end{pmatrix}$ and $\mathbf{P}^2 + 3\mathbf{P} + \mathbf{Q} = \begin{pmatrix} 0 & 0 \\ 0 & 0 \end{pmatrix}$.

23 $\mathbf{R} = \begin{pmatrix} 3 & 7 \\ 2 & 5 \end{pmatrix}$ and $\mathbf{S} = \begin{pmatrix} -3 & 2 \\ 4 & -1 \end{pmatrix}$. Find the matrix $\mathbf{M}$ if: (a) $2\mathbf{M} = \mathbf{R} + \mathbf{S}$, (b) $\mathbf{M} + \mathbf{R} = \begin{pmatrix} 0 & 0 \\ 0 & 0 \end{pmatrix}$,
(c) $\mathbf{MR} = \begin{pmatrix} 1 & 0 \\ 0 & 1 \end{pmatrix}$, (d) $\mathbf{M} + \mathbf{R} = 4\mathbf{S}$, (e) $5\mathbf{M} + 2\mathbf{S} = \mathbf{R}$.
How many solutions are there to the equation $\mathbf{MR} = \mathbf{RM}$?

24 $\mathbf{P} = \begin{pmatrix} 5 & 7 \\ 2 & 3 \end{pmatrix}$. Find matrix $\mathbf{M}$ if:
(a) $\mathbf{MP} = \begin{pmatrix} 7 & 10 \\ 7 & 10 \end{pmatrix}$, (b) $\mathbf{MP} = \begin{pmatrix} 7 & 10 \\ 14 & 20 \end{pmatrix}$, (c) $\mathbf{MP} = \begin{pmatrix} 5 & 7 \\ 0 & 0 \end{pmatrix}$,
(d) $\mathbf{MP} = \begin{pmatrix} 0 & 0 \\ 4 & 6 \end{pmatrix}$, (e) $\mathbf{MP} = \begin{pmatrix} 1 & 0 \\ 0 & 1 \end{pmatrix}$, (f) $\mathbf{PM} = \begin{pmatrix} 1 & 0 \\ 0 & 1 \end{pmatrix}$

25 $\mathbf{Q} = \begin{pmatrix} 4 & 1 \\ 7 & 2 \end{pmatrix}$. Find matrix $\mathbf{N}$ if $\mathbf{NQ}$ equals:
(a) $\begin{pmatrix} 11 & 3 \\ 11 & 3 \end{pmatrix}$, (b) $\begin{pmatrix} 22 & 6 \\ 11 & 3 \end{pmatrix}$, (c) $\begin{pmatrix} 4 & 1 \\ 0 & 0 \end{pmatrix}$, (d) $\begin{pmatrix} 1 & 0 \\ 0 & 1 \end{pmatrix}$,
(e) $\begin{pmatrix} -2 & 0 \\ 0 & -2 \end{pmatrix}$.

26 (a) Calculate $\begin{pmatrix} r & s \\ s & r \end{pmatrix}^2$. (b) Find matrix $\mathbf{Q}$ if $\mathbf{Q}^2$ equals:
(1) $\begin{pmatrix} 5 & 4 \\ 4 & 5 \end{pmatrix}$ (2) $\begin{pmatrix} 17 & 8 \\ 8 & 17 \end{pmatrix}$ (3) $\begin{pmatrix} 20 & 16 \\ 16 & 20 \end{pmatrix}$ (4) $\begin{pmatrix} 10 & 6 \\ 6 & 10 \end{pmatrix}$.

27 Solve the simultaneous equations:

(a) $5x + 3y = 9$, $13x + 8y = 22$
(b) $9x + 5y = 35$, $16x + 9y = 61$
(c) $23x - 14y = 22$, $5x - 3y = 5$
(d) $11x + 2y = 4$, $10x + 4y = -16$
(e) $6x + 7y = 23$, $8x + 9y = 33$
(f) $2x + 5y = 31$, $3x - 6y = 6$
(g) $4x + 3y = 14$, $5x + 7y = -2$
(h) $5x + 3y = 12$, $4x + 5y = -6$.

28

$$\mathbf{D} = \begin{array}{c} \\ m \\ t \\ w \\ th \\ f \end{array} \begin{array}{c} \text{Tea} \quad \text{Coffee} \quad \text{Chocolate} \\ \begin{pmatrix} 33 & 42 & 55 \\ 28 & 35 & 43 \\ 56 & 64 & 41 \\ 36 & 49 & 38 \\ 41 & 53 & 28 \end{pmatrix} \end{array} \qquad \mathbf{C} = \begin{array}{c} \text{Tea} \\ \text{Cof.} \\ \text{Choc.} \end{array} \begin{pmatrix} 2p \\ 3p \\ 3p \end{pmatrix}$$

Matrix **D** shows the daily sales of drinks from a Hot Drinks Machine for each of the 5 days of one week.

Matrix **C** shows the cost of each type of drink.

(a) Calculate $(1\ 1\ 1\ 1\ 1)\mathbf{D}$ and say what information this gives.

(b) Use this result and matrix C to find the week's takings.

(c) Calculate $\mathbf{D}\begin{pmatrix} 1 \\ 1 \\ 1 \end{pmatrix}$ and say what information this gives.

(d) Find $(1\ 1\ 1\ 1\ 1)\mathbf{D}\begin{pmatrix} 1 \\ 1 \\ 1 \end{pmatrix}$. What does this represent?

29 $$\mathbf{K} = \begin{array}{c} \\ S_1 \\ S_2 \\ S_3 \end{array} \begin{array}{c} R_1\ R_2\ R_3\ R_4 \\ \begin{pmatrix} 6 & 8 & 3 & 4 \\ 5 & 7 & 4 & 5 \\ 8 & 3 & 5 & 1 \end{pmatrix} \end{array} \qquad \mathbf{V} = \begin{array}{c} R_1 \\ R_2 \\ R_3 \\ R_4 \end{array} \begin{pmatrix} 18 \\ 20 \\ 24 \\ 35 \end{pmatrix}$$

$$\mathbf{W} = \begin{array}{c} \\ S_1 \\ S_2 \\ S_3 \end{array} \begin{array}{c} R_1\ R_2\ R_3\ R_4 \\ \begin{pmatrix} 2 & 2 & 1 & 3 \\ 1 & 3 & 1 & 2 \\ 3 & 1 & 2 & 0 \end{pmatrix} \end{array}$$

Matrix **K** shows the stock of four types of record players R_1, R_2, R_3 and R_4 in three shops S_1, S_2 and S_3. Matrix **V** shows the value of the record players in £'s. Matrix **W** gives the week's sales. Find:

(a) the stock at the end of the week, (b) the order matrix to bring the stock of each of the cheaper pair of players to 8 and the dearer pair to 5, (c) the value of the sales, (d) the value of the order.

30 $$\mathbf{S} = \begin{array}{c} \\ C_1 \\ C_2 \\ C_3 \end{array} \begin{array}{c} S_1\ S_2\ S_3\ S_4 \\ \begin{pmatrix} 2 & 1 & 3 & 0 \\ 4 & 2 & 1 & 4 \\ 3 & 2 & 1 & 4 \end{pmatrix} \end{array} \qquad \mathbf{D} = \begin{pmatrix} 3 & 4 & 2 & 5 \\ 1 & 3 & 4 & 3 \\ 2 & 3 & 4 & 1 \end{pmatrix}$$

$$\mathbf{L} = \begin{pmatrix} 2 & 1 & 2 & 4 \\ 1 & 3 & 1 & 2 \\ 2 & 3 & 4 & 2 \end{pmatrix}$$

Matrix **S** Shows the stock of 3 types of cooker C_1, C_2 and C_3 in 4 showrooms S_1, S_2, S_3 and S_4. Matrix **D** shows the deliveries of new cookers at the beginning of a week. Matrix **L** shows the sales during that week.

Find (a) the stock immediately after delivery **D**, (b) the stock at the end of the week, (c) the order matrix to bring stocks of all cookers in all showrooms up to 6.

31 $\mathbf{V} = \overset{\begin{matrix} C_1 & C_2 & C_3 \end{matrix}}{(45 \quad 30 \quad 40)}$ and $\mathbf{P} = \overset{\begin{matrix} C_1 & C_2 & C_3 \end{matrix}}{(8 \quad 5 \quad 6)}$ give the value of each cooker and the profit on each cooker in £'s, in question 30.

Calculate (a) the value of the original stock, (b) the profit on the week's sales, (c) the cost of the order.

32

$$\mathbf{W} = \begin{matrix} & P_1 & P_2 & P_3 \\ W_1 \\ W_2 \\ W_3 \end{matrix}\!\!\begin{matrix} \\ \begin{pmatrix} 5 & 3 & 4 \\ 4 & 6 & 2 \\ 1 & 5 & 4 \end{pmatrix} \end{matrix} \quad \mathbf{L} = \begin{matrix} P_1 \\ P_2 \\ P_3 \end{matrix}\begin{pmatrix} 5l \\ 3l \\ 4l \end{pmatrix} \quad \mathbf{C} = \begin{matrix} P_1 \\ P_2 \\ P_3 \end{matrix}\begin{pmatrix} 1\text{p} \\ 3\text{p} \\ 2\text{p} \end{pmatrix}$$

A laundry has 3 types of wash W_1, W_2 and W_3 each involving 3 processes P_1, P_2 and P_3. The time for each process, in minutes, is given by table **W**. The amount of water used, in litres per minute, is given by table **L**. The cost, in pence per minute, for each process is given by table **C**.

(a) From matrices **W** and **L** calculate the amount of water used in each wash. (b) Calculate the cost of each wash. (c) What information is given by $\mathbf{W}\begin{pmatrix} 1 \\ 1 \\ 1 \end{pmatrix}$?

33

$$\mathbf{M} = \begin{matrix} & T_1 & T_2 & T_3 \\ B_1 \\ B_2 \\ B_3 \end{matrix}\!\!\begin{matrix} \\ \begin{pmatrix} 3 & 2 & 1 \\ 4 & 1 & 3 \\ 3 & 5 & 2 \end{pmatrix} \end{matrix} \quad \mathbf{C} = \begin{matrix} T_1 \\ T_2 \\ T_3 \end{matrix}\begin{pmatrix} 40\text{p} \\ 44\text{p} \\ 36\text{p} \end{pmatrix}$$

Matrix **M** gives the proportions of three teas T_1, T_2 and T_3 in three blends of tea B_1, B_2 and B_3. Matrix **C** gives the cost per kg of each tea.

(a) What information is given by $\mathbf{M}\begin{pmatrix} 1 \\ 1 \\ 1 \end{pmatrix}$?

(b) Calculate **MC** and hence the cost per kg of each blend to the nearest p.

Matrices: Transformations

Exercise 18

1 Set up an x and a y axis using scales of 1 cm to 1 unit for both. (a) Plot the points $\mathbf{T} = \begin{matrix} x \\ y \end{matrix}\begin{pmatrix} 2 & 2 & 1 & 3 \\ 0 & 2 & 2 & 2 \end{pmatrix}$ and join these to form the letter T.

(b) Calculate and plot $\mathbf{T} + \begin{pmatrix} 3 & 3 & 3 & 3 \\ 4 & 4 & 4 & 4 \end{pmatrix}$ and $\mathbf{T} + \begin{pmatrix} -4 & -4 & -4 & -4 \\ 2 & 2 & 2 & 2 \end{pmatrix}$.

(c) What transformations are these? Say how far the T moves parallel to each axis in the two cases. (d) What matrix is needed to restore the T to its original position in the two cases? (e) What matrix would move the T 5 along and 3 up?

2 (a) Using axes as in question 1, plot the points $\begin{matrix} x \\ y \end{matrix}\begin{pmatrix} 1 & 2 & 3 \\ 2 & 4 & 4 \end{pmatrix}$ and join these to form a triangle.

(b) $\mathbf{T}_1 = \begin{pmatrix} 2 & 2 & 2 \\ 3 & 3 & 3 \end{pmatrix}$ and $\mathbf{T}_2 = \begin{pmatrix} 3 & 3 & 3 \\ -5 & -5 & -5 \end{pmatrix}$ are two translations. Plot the images of the triangle under (1) $\mathbf{T}_1$ followed by $\mathbf{T}_2$ and (2) $\mathbf{T}_2$ followed by $\mathbf{T}_1$. Comment on your results.

(c) To what single translation are the above equivalent?

3 $\mathbf{T} = \begin{matrix} x \\ y \end{matrix}\begin{pmatrix} 1 & 1 & 2 \\ 1 & 4 & 3 \end{pmatrix}$ gives three points forming a triangle.

$\mathbf{A} = \begin{pmatrix} 3 & 3 & 3 \\ 2 & 2 & 2 \end{pmatrix}$ $\mathbf{B} = \begin{pmatrix} -2 & -2 & -2 \\ 3 & 3 & 3 \end{pmatrix}$ and

$\mathbf{C} = \begin{pmatrix} 2 & 2 & 2 \\ -4 & -4 & -4 \end{pmatrix}$ are three translations.

Plot **T** and the images of **T** under the translations (a) **A** followed by **B** followed by **C**, (b) **B** followed by **C** followed by **A**, (c) **C** followed by **A** followed by **B**.

Comment on your results and state the single translation to which these are equivalent.

4 (a) Using 1 cm to 1 unit for both axes, plot the triangle

$\mathbf{G} = \begin{matrix} x \\ y \end{matrix}\begin{pmatrix} 1 & 3 & 2 \\ 3 & 3 & 1 \end{pmatrix}$. (b) Calculate and plot $\begin{pmatrix} 1 & 0 \\ 0 & -1 \end{pmatrix}\mathbf{G}$, $\begin{pmatrix} -1 & 0 \\ 0 & 1 \end{pmatrix}\mathbf{G}$, $\begin{pmatrix} -1 & 0 \\ 0 & -1 \end{pmatrix}\mathbf{G}$.

(c) In each case state in what axis or point the triangle has been reflected. (d) What is the relation between the first two reflections and the third?

5 Label the original triangle in question 4, ABC clockwise. Now label the corresponding points on the three images. Which of these are clockwise and which anticlockwise? What is the effect of one reflection on the labelling of the triangle? What is the effect of two reflections. Can you extend these results?

6 (a) Using 1 cm to 1 unit on both axes, plot the triangle

$$\begin{matrix} & \begin{matrix} R & S & T \end{matrix} \\ \begin{matrix} x \\ y \end{matrix} & \begin{pmatrix} 1 & 5 & 2 \\ 4 & 5 & 3 \end{pmatrix} \end{matrix}.$$

(b) Plot the image of the triangle when its matrix is premultiplied by $\begin{pmatrix} 0 & 1 \\ 1 & 0 \end{pmatrix}$. What transformation is this?

(c) Say whether the following are changed or unchanged under the transformation: (1) the area, (2) the length RT, (3) the angle RST, (4) the direction RS, (5) the clockwise labelling of RST.

7 (a) With scales as before plot the rectangle $\begin{matrix} & \begin{matrix} A & B & C & D \end{matrix} \\ \begin{matrix} x \\ y \end{matrix} & \begin{pmatrix} 2 & 2 & 4 & 4 \\ 1 & 2 & 2 & 1 \end{pmatrix} \end{matrix}$.

Shade the triangle ABC.

(b) Plot the images of the rectangle when its matrix is premultiplied by $\begin{pmatrix} 0 & -1 \\ 1 & 0 \end{pmatrix}$, $\begin{pmatrix} -1 & 0 \\ 0 & -1 \end{pmatrix}$, $\begin{pmatrix} 0 & 1 \\ -1 & 0 \end{pmatrix}$.

(c) Explain the effect of these transformations.

8 Using 1 cm to 1 unit for both axes, plot the triangle $\mathbf{T} = \begin{matrix} x \\ y \end{matrix}\begin{pmatrix} 1 & 1 & 4 \\ 1 & 2 & 2 \end{pmatrix}$. If $\mathbf{R} = \begin{pmatrix} 0 & -1 \\ 1 & 0 \end{pmatrix}$, plot $\mathbf{RT}$, $\mathbf{R}^2\mathbf{T}$, $\mathbf{R}^3\mathbf{T}$.

Explain the effect of premultiplying by the matrix **R**. What is the effect of $\mathbf{R}^6\mathbf{T}$?

9 (a) With scales as before, plot the points $\begin{matrix} x \\ \end{matrix}\begin{pmatrix} 1 & 1\frac{1}{2} & 2 \\ 2 & 1 & 2 \end{pmatrix}$ and join these to form a letter V.

(b) Plot the image of the V when its matrix is premultiplied by (1) $\begin{pmatrix} 2 & 0 \\ 0 & 2 \end{pmatrix}$, (2) $\begin{pmatrix} 3 & 0 \\ 0 & 3 \end{pmatrix}$, (3) $\begin{pmatrix} -4 & 0 \\ 0 & -4 \end{pmatrix}$.

(c) Explain the effect of these transformations.

(d) What matrix would transform the image in case 2 into the image of case 3?

10 (a) With scales of 1 cm to 1 unit for both axes plot the points $\begin{matrix}x\\y\end{matrix}\begin{pmatrix}1 & 1 & 1 & 1\frac{1}{2} & 2\\1 & 2 & 3 & 2 & 3\end{pmatrix}$ and join them to form the letter F.

(b) Plot $\begin{pmatrix}3 & 0\\0 & 3\end{pmatrix}\mathbf{F}$ and $\begin{pmatrix}-2 & 0\\0 & -2\end{pmatrix}\mathbf{F}$.

(c) Join the corresponding points on the three F's and comment on the result.

11 (a) Plot triangle $\mathbf{T} = \begin{matrix}x\\y\end{matrix}\begin{pmatrix}1 & 3 & 2\\3 & 3 & 1\end{pmatrix}$.

(b) Calculate and plot $\begin{pmatrix}2 & 0\\0 & 2\end{pmatrix}\mathbf{T}$ and $\begin{pmatrix}\frac{1}{2} & 0\\0 & \frac{1}{2}\end{pmatrix}\mathbf{T}$.

(c) What transformations are these? What is the ratio of (1) the new lengths to the old, (2) the new areas to the old?

(d) Say whether the following vary or are invariant under the above transformations. (1) angles, (2) lengths, (3) directions, (4) areas, (5) shape.

12 Draw the rectangle **L** with vertices $\begin{matrix}x\\y\end{matrix}\begin{pmatrix}1 & 1 & 3 & 3\\1 & 2 & 2 & 1\end{pmatrix}$.

Calculate and plot (a) $\begin{pmatrix}3 & 0\\0 & 3\end{pmatrix}\mathbf{L}$, (b) $\begin{pmatrix}-2 & 0\\0 & -2\end{pmatrix}\mathbf{L}$, (c) $\begin{pmatrix}\frac{1}{2} & 0\\0 & \frac{1}{2}\end{pmatrix}\mathbf{L}$.

For each transformation write down the ratio of (1) the new lengths to the old, (2) the new areas to the old.

13 On axes with scales of 1 cm to 1 unit, plot the points $\mathbf{N} = \begin{matrix}x\\y\end{matrix}\begin{pmatrix}1 & 1 & 3 & 3\\1 & 3 & 1 & 3\end{pmatrix}$ and join these to form the letter N.

Calculate and plot $\begin{pmatrix}2 & 0\\0 & 1\end{pmatrix}\mathbf{N}$ and $\begin{pmatrix}1 & 0\\0 & 2\end{pmatrix}\mathbf{N}$.

What transformations are these?

14 Plot the 'dog' in Fig. 18.1, using the matrix $\mathbf{D} = \begin{matrix}x\\y\end{matrix}\begin{pmatrix}0 & 1 & 1 & 3 & 3 & 3 & 4\\2 & 0 & 1 & 0 & 1 & 2 & 2\end{pmatrix}$.

Calculate and plot $\begin{pmatrix}1 & 0\\0 & 2\end{pmatrix}\mathbf{D}$ and $\begin{pmatrix}2 & 0\\0 & 1\end{pmatrix}\mathbf{D}$.

Comment on the changes in the dog.

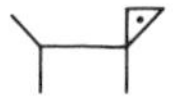

Fig. 18.1

15 Plot $\begin{pmatrix}1 & 2\\0 & 1\end{pmatrix}\mathbf{D}$ for the dog in question 14.

What transformation is this?

16 Plot the square $\mathbf{S} = \begin{matrix} x \\ y \end{matrix}\begin{pmatrix} 1 & 1 & 3 & 3 \\ 0 & 2 & 0 & 2 \end{pmatrix}$.

Plot the image of the square under (a) the stretch $\begin{pmatrix} 2 & 0 \\ 0 & 1 \end{pmatrix}$, (b) the shear $\begin{pmatrix} 1 & 0 \\ 2 & 1 \end{pmatrix}$, (c) the enlargement $\begin{pmatrix} -2 & 0 \\ 0 & -2 \end{pmatrix}$.

In each case calculate the ratio of the new area to the old.

17 The determinant of a matrix $\begin{pmatrix} a & b \\ c & d \end{pmatrix}$ is written $\begin{vmatrix} a & b \\ c & d \end{vmatrix}$ and the value of $\begin{vmatrix} a & b \\ c & d \end{vmatrix}$ is ad − bc.

For example $\begin{vmatrix} 4 & 5 \\ 2 & 3 \end{vmatrix} = 12 - 10 = 2$.

Calculate the value of $\begin{vmatrix} 2 & 0 \\ 0 & 1 \end{vmatrix}$, $\begin{vmatrix} 1 & 0 \\ 2 & 1 \end{vmatrix}$ and $\begin{vmatrix} 2 & 0 \\ 0 & 2 \end{vmatrix}$.

Compare these answers with the ratios of the areas in question 16. Comment on your results.

18 Calculate the ratio of the new areas to the old when pre-multiplying by $\begin{pmatrix} 4 & 3 \\ 2 & 4 \end{pmatrix}$, $\begin{pmatrix} 3 & 1 \\ 0 & 1 \end{pmatrix}$, $\begin{pmatrix} 5 & 1 \\ 2 & 1 \end{pmatrix}$, $\begin{pmatrix} 4 & 0 \\ 0 & 4 \end{pmatrix}$, $\begin{pmatrix} 6 & 2 \\ 1 & 2 \end{pmatrix}$.

19 Set up a positive x and y axis for values from 0 to 16. Plot the square $\mathbf{S} = \begin{matrix} x \\ y \end{matrix}\begin{pmatrix} 1 & 1 & 2 & 2 \\ 1 & 2 & 2 & 1 \end{pmatrix}$.

Plot $\begin{pmatrix} 2 & 4 \\ 1 & 2 \end{pmatrix}\mathbf{S}$. What special feature is there about the image of the square? What happens to its area?

Calculate $\begin{vmatrix} 2 & 4 \\ 1 & 2 \end{vmatrix}$ and comment.

20 Repeat question 19 for $\begin{pmatrix} 3 & 1 \\ 6 & 2 \end{pmatrix}\mathbf{S}$.

21 Plot another square **S**, as in question 19. Calculate and plot (a) $\begin{pmatrix} 3 & 0 \\ 0 & 3 \end{pmatrix}\mathbf{S}$, (b) $\begin{pmatrix} 3 & 0 \\ 0 & 1 \end{pmatrix}\mathbf{S}$, (c) $\begin{pmatrix} 1 & 0 \\ 1 & 3 \end{pmatrix}\mathbf{S}$.

Name each transformation. Calculate the ratio of the new areas to the old in each case. Compare your answers with the values of the three determinants.

22 What are the three matrices that would restore the images in question 21 to the original square **S**?

23 Plot the points $\mathbf{W} = \begin{pmatrix} 1 & 2 & 3 & 4 & 5 \\ 3 & 1 & 3 & 1 & 3 \end{pmatrix}$ and join these to form

the letter W. Plot the images of **W** when premultiplied by (a) $\begin{pmatrix} -1 & 0 \\ 0 & 1 \end{pmatrix}$, (b) $\begin{pmatrix} 0 & -1 \\ 1 & 0 \end{pmatrix}$, (c) $\begin{pmatrix} -\frac{1}{2} & 0 \\ 0 & -\frac{1}{2} \end{pmatrix}$.

Name these transformations. Say what matrix operators should be used to restore the images to the original **W**.

24 (a) What transformation is effected by the matrix $\mathbf{P} = \begin{pmatrix} 1 & 0 \\ 0 & -1 \end{pmatrix}$, $\mathbf{Q} = \begin{pmatrix} 0 & -1 \\ 1 & 0 \end{pmatrix}$, $\mathbf{R} = \begin{pmatrix} 4 & 0 \\ 0 & 4 \end{pmatrix}$?

(b) In which are (1) the areas, (2) the directions, (3) the shapes, unchanged?

25 Which of the following matrix operators (a) enlarges, (b) rotates, (c) reflects?
$\mathbf{W} = \begin{pmatrix} 1 & 0 \\ 0 & -1 \end{pmatrix}$, $\mathbf{X} = \begin{pmatrix} -2 & 0 \\ 0 & 2 \end{pmatrix}$, $\mathbf{Y} = \begin{pmatrix} 0 & 1 \\ -1 & 0 \end{pmatrix}$,
$\mathbf{Z} = \begin{pmatrix} 0 & -3 \\ 3 & 0 \end{pmatrix}$.

Does $\mathbf{WX} = \mathbf{XW}$?

26 Write down the matrix operator that produces (a) a reflection in the x-axis, (b) a reflection in the y-axis, (c) an anticlockwise rotation of 90°, (d) a reflection in the x-axis and all lengths doubled, (e) a reflection in the y-axis and areas increased 9 times.

Matrices: Networks and Relations

Exercise 19

Networks

1 These questions refer to the network in Fig. 19.1. (a) Form the matrix **R** giving the one-stage routes for the network. (b) What information is given by the row totals and by the

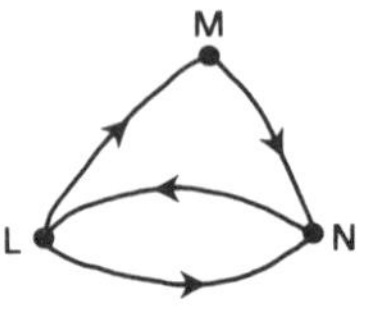

Fig. 19.1

column totals? (c) Explain why the elements in the leading diagonal are all zeros. (d) Calculate $\mathbf{R}^2$. What is the total number of two-stage routes? (e) How many three-stage routes are there? How many of these begin and end at the same point?

2 Refer to the network in Fig. 19.2. (a) Write out a matrix **R** for the one-stage routes on this network. What is the total

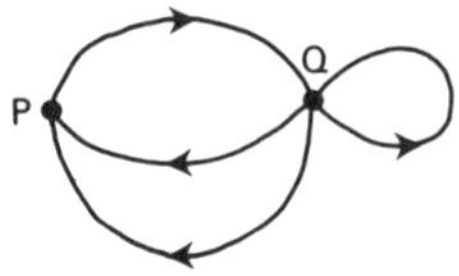

Fig. 19.2

number of such routes? (b) Calculate $\mathbf{R}^2$ and $\mathbf{R}^3$. (c) How many two-stage routes are there? (d) List the three-stage routes that begin at **P** and end at **Q**.

3 Write out the matrix **R** for one-stage routes for the network in Fig. 19.3. Calculate $\mathbf{R}^2$ and $\mathbf{R}^4$.

State the total number of 1-stage, 2-stage and 4-stage routes. How many 4-stage routes begin and end at the same point?

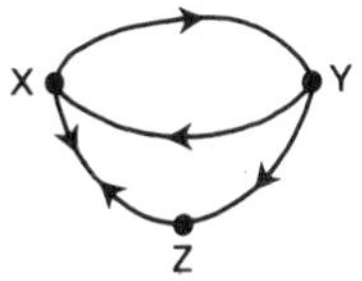

Fig. 19.3

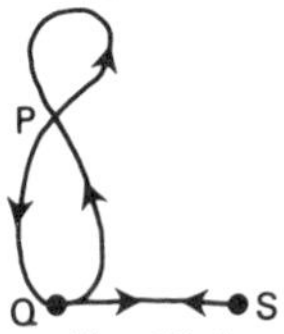

Fig. 19.4

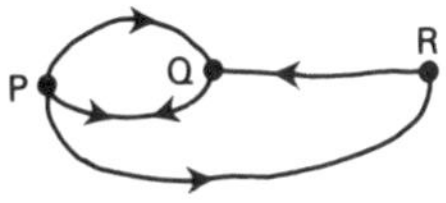

Fig. 19.5

4 Find the total number of 1, 2 and 3-stage routes for the network in Fig. 19.4. List the 3-stage routes from Q to S.

5 Calculate the number of 1, 2 and 3-stage routes for the network in Fig. 19.5.

How many 3-stage routes are there (a) beginning at P, (b) ending at Q, (c) beginning and ending at R?

For which pairs of points are there no 3-stage routes?

6 Form matrices **R**, $\mathbf{R}^2$ and $\mathbf{R}^3$ to show the 1-stage, 2-stage and 3-stage routes for the network in Fig. 19.6.

Write down $\mathbf{R}^4$ and $\mathbf{R}^6$ with the minimum further calculation. How many 6-stage routes are there?

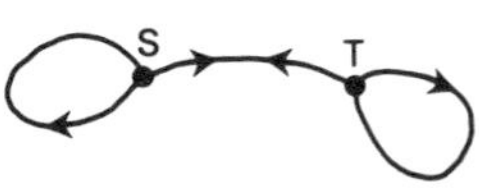

Fig. 19.6

7 LMS is an old railway line (Fig. 19.7). A single track runs from L to M and from M to S. A double connects S and L. Form matrices **R**, $\mathbf{R}^2$ and $\mathbf{R}^4$ for the 1, 2 and 4-stage routes for L, M and S.

How many 2-stage routes are there? How many 4-stage routes begin and end at different points? Which point has no 1, 2 or 4-stage route beginning and ending at it? Is there a 3-stage route beginning and ending at that point?

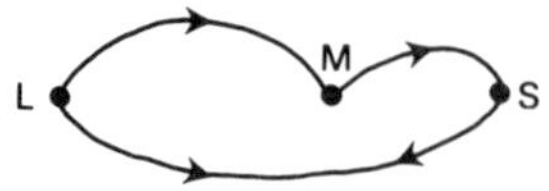

Fig. 19.7

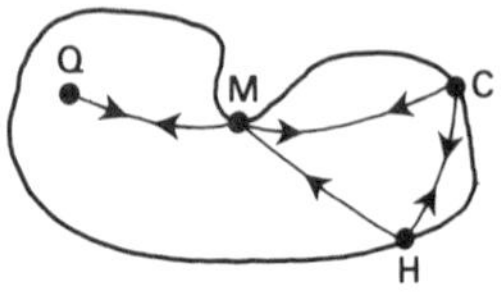

Fig. 19.8

8 Figure 19.8 shows an island with roads connecting Mainport M, Oyster Creek C, Lobster Head H and the Quarries Q. The road from H to M is narrow and traffic is allowed in the direction HM only.

Form matrices **R** and $\mathbf{R}^2$ to show the one and two-stage routes on the island. Which places – if any – are connected neither by a one nor a two-stage route?

Suggest a route to deliver stone from the quarry to C and H and oysters and lobsters to M.

A landslide completely blocks the road MC and it will take 3 weeks to clear. Suggest how road communications should be maintained on the island.

Relations.

9 (a) Write a matrix **C** for the relation 'is 200 m higher than' for the set of contours P, Q, R and S in Fig. 19.9.

(b) What does the transpose of matrix **C** represent?

(c) Calculate matrix $\mathbf{C}^2$ and say what relation this gives. What relation would the matrix $\mathbf{C}^3$ represent?

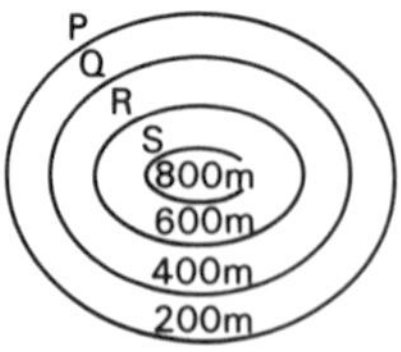

Fig. 19.9

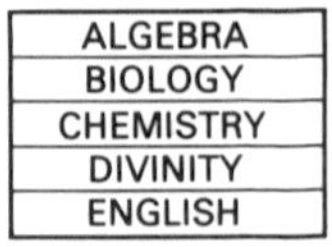

Fig. 19.10

10 Figure 19.10 shows the arrangement of sets of books on a bookshelf. (a) Form a matrix **M** for the relation 'is below' for the sets of books A, B, C, D and E. (b) Form the transpose of **M**. What relation does this give? (c) Calculate $\mathbf{M}^2$. What meaning can be attached to $\mathbf{M}^2$?

11 A matrix **T** represents the relation 'is three times as long as' for a set of pieces of string of different lengths. What relation is represented by (a) the transpose of **T**, (b) $\mathbf{T}^2$?

12 Mr Kaye, Mr Lewis, Mr Marsden and Mr Newt live in a terrace of four cottages as shown below.

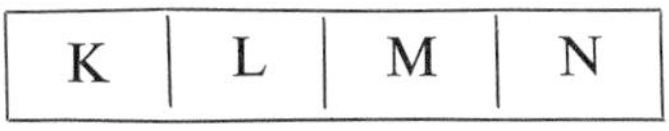

(a) Write out a matrix **B** for the relation 'is a next door neighbour of' on {K L M N}. (b) Suggest why the numbers in the leading diagonal are zero. (c) Form the transpose of **B**. What do you notice about this? Suggest a reason.

13 Messrs North, South, East and West sit around a table as shown in Fig. 19.11.

(a) Write a matrix **R** for the relation 'is on the right of' for N, S, E, W. (b) What relation would the transpose of **R** represent? (c) Calculate $\mathbf{R}^2$ and suggest a meaning for this. Explain why matrix $\mathbf{R}^2$ is symmetrical.

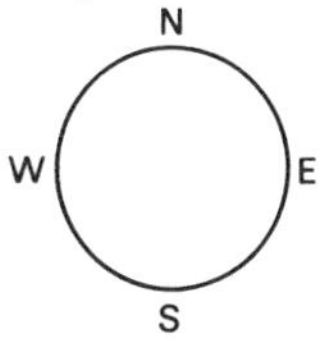

Fig. 19.11

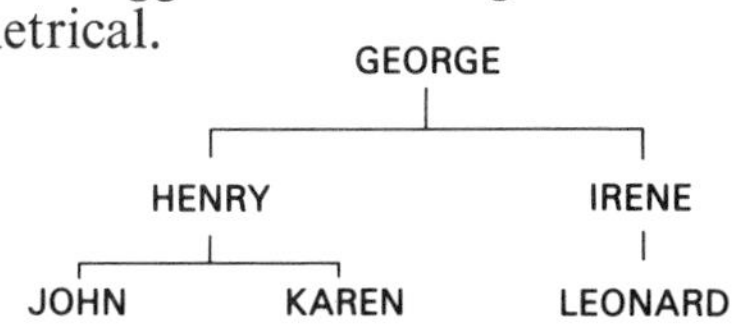

Fig. 19.12

14 Consider the family tree in Fig. 19.12. (a) Write out the matrix **F** for the relation 'is the father of'. (b) Write down the transpose of **F** and say what relation this represents. (c) Suggest a meaning for $\mathbf{F}^2$.

15 For the family tree in Fig. 19.12, write out a matrix **B** for the relation 'is the brother of' and a matrix **P** for the relation 'is a parent of'. Form the product BP. What relation does this represent for the elements given?

16 Figure 19.13. shows 'mountains' W, X, Y and Z. (a) Form the matrix **H** for the relation 'is higher than' on {W X Y Z}. (b) What relation does the transpose of **H** represent? (c) Calculate $\mathbf{H}^2$ and suggest a meaning for this matrix.

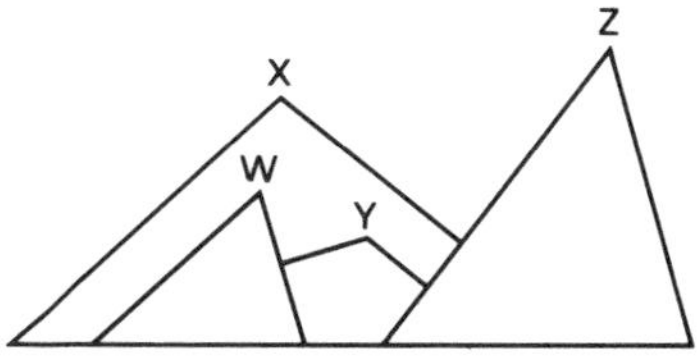

Fig. 19.13

17 (a) Write a matrix **T** for the relation 'is 2 more than' on

{2 4 6 16 18}. (b) Write out the matrix **S** for 'is the square of' on the same set. (c) Calculate **ST** and **TS**. (d) Suggest what relations these represent for the given elements.

18 Peter, Quentin, Ralph, Sarah, Terry and Una sit around a hexagonal table as shown in Fig. 19.14. (a) Form the matrix **L** to show the relation 'sits to the left of' on {P Q R S T U}. (b) Suggest a meaning for the transpose of **L**. (c) Calculate $\mathbf{L}^2$ and $\mathbf{L}^3$. (d) Which one of these represents 'sits opposite'? What does the other represent? (e) Why should $\mathbf{L}^5$ equal the transpose of **L**?

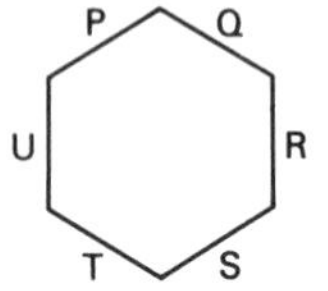

Fig. 19.14

Vectors

Exercise 20

1 Make a copy of Fig. 20.1. Mark with a different colour each set of equivalent vectors.

Express these sets in the form $\begin{pmatrix} x \\ y \end{pmatrix}$, taking 1 square to 1 unit.

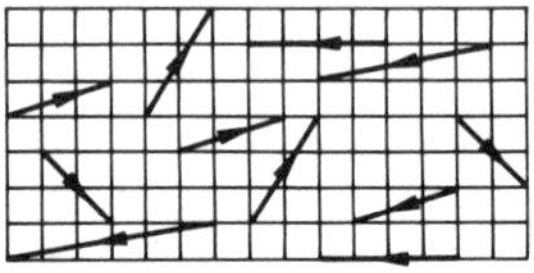

Fig. 20.1

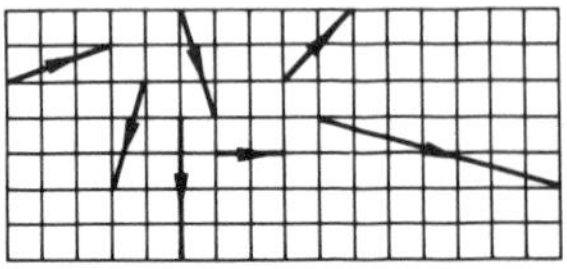

Fig. 20.2

2 Express in the form $\begin{pmatrix} x \\ y \end{pmatrix}$ the vectors illustrated in Fig. 20.2. Take 1 square as 1 unit.

3 (a) On squared paper draw diagrams to illustrate $\mathbf{p} = \begin{pmatrix} 4 \\ 2 \end{pmatrix}$, $\mathbf{q} = \begin{pmatrix} 3 \\ 4 \end{pmatrix}$ and $\mathbf{r} = \begin{pmatrix} -4 \\ 2 \end{pmatrix}$.

(b) Draw further diagrams to illustrate $\mathbf{p} + \mathbf{q}$, $\mathbf{q} + \mathbf{r}$, $3\mathbf{q}$, $2\mathbf{p} + 3\mathbf{q}$, $\mathbf{p} - \mathbf{r}$, $\mathbf{p} + \mathbf{q} - \mathbf{r}$.

(c) State the resultants of the vectors in part (b).

4 Repeat question 3 for $\mathbf{p} = \begin{pmatrix} -1 \\ 3 \end{pmatrix}$, $\mathbf{q} = \begin{pmatrix} 4 \\ 2 \end{pmatrix}$, $\mathbf{r} = \begin{pmatrix} 5 \\ -4 \end{pmatrix}$.

5 State the vector that translates the point P (3,2) to the point (a) (6,4), (b) (5,1), (c) (1,5), (d) (0,0), (e) (−3,4), (f) (−6,−4).

6 Find by drawing the magnitude (modulus) and the direction, relative to the positive x-axis, of the following vectors:
$\begin{pmatrix}4\\6\end{pmatrix}$ $\begin{pmatrix}3\\7\end{pmatrix}$ $\begin{pmatrix}8\\4\end{pmatrix}$ $\begin{pmatrix}8\\-4\end{pmatrix}$ $\begin{pmatrix}-5\\6\end{pmatrix}$ $\begin{pmatrix}-5\\-6\end{pmatrix}$ $\begin{pmatrix}7\\-2\end{pmatrix}$ $\begin{pmatrix}-9\\-5\end{pmatrix}$.

7 Calculate the modulus and the direction, relative to the positive x-axis of the vectors:
$\begin{pmatrix}3\\4\end{pmatrix}$ $\begin{pmatrix}6\\8\end{pmatrix}$ $\begin{pmatrix}6\\-8\end{pmatrix}$ $\begin{pmatrix}9\\12\end{pmatrix}$ $\begin{pmatrix}5\\12\end{pmatrix}$ $\begin{pmatrix}5\\-12\end{pmatrix}$ $\begin{pmatrix}-5\\12\end{pmatrix}$ $\begin{pmatrix}10\\24\end{pmatrix}$
$\begin{pmatrix}15\\8\end{pmatrix}$ $\begin{pmatrix}8\\15\end{pmatrix}$.

8 Calculate the magnitude of the resultants of
(a) $\begin{pmatrix}1\\6\end{pmatrix}+\begin{pmatrix}2\\-2\end{pmatrix}$, (b) $\begin{pmatrix}16\\10\end{pmatrix}-\begin{pmatrix}4\\5\end{pmatrix}$, (c) $\begin{pmatrix}9\\3\end{pmatrix}+\begin{pmatrix}-3\\5\end{pmatrix}$,
(d) $\begin{pmatrix}8\\15\end{pmatrix}+\begin{pmatrix}12\\6\end{pmatrix}$.

9 Solve the vector equations:
(a) $\begin{pmatrix}x\\6\end{pmatrix}+\begin{pmatrix}3\\y\end{pmatrix}=\begin{pmatrix}8\\2\end{pmatrix}$, (b) $\begin{pmatrix}p\\7\end{pmatrix}+\begin{pmatrix}-4\\q\end{pmatrix}=\begin{pmatrix}5\\5\end{pmatrix}$.

10 Find p and q if
(a) $\begin{pmatrix}p\\q\end{pmatrix}+3\begin{pmatrix}p\\-q\end{pmatrix}=\begin{pmatrix}12\\-6\end{pmatrix}$, (b) $\begin{pmatrix}p\\-q\end{pmatrix}+\begin{pmatrix}q\\p\end{pmatrix}=\begin{pmatrix}9\\3\end{pmatrix}$.

11 Find x if the displacements represented by
(a) $\begin{pmatrix}x\\4\end{pmatrix}$ and $\begin{pmatrix}6\\8\end{pmatrix}$ are parallel, (b) $\begin{pmatrix}x\\5\end{pmatrix}$ and $\begin{pmatrix}20\\x\end{pmatrix}$ are parallel, (c) $\begin{pmatrix}x\\9\end{pmatrix}$ and $\begin{pmatrix}6\\7\end{pmatrix}$ are equal in length.

12 $\mathbf{r}=\begin{pmatrix}2\\1\\3\end{pmatrix}$ and $\mathbf{s}=\begin{pmatrix}3\\-2\\4\end{pmatrix}$. Express in component form:
(a) $-\mathbf{r}$, (b) $\mathbf{r}+\mathbf{s}$, (c) $\mathbf{r}-\mathbf{s}$, (d) $\mathbf{s}-\mathbf{r}$, (e) $\frac{1}{2}(\mathbf{r}+\mathbf{s})$, (f) $\frac{1}{2}(\mathbf{r}-\mathbf{s})$.

13 $\mathbf{p}=\begin{pmatrix}2\\2\\3\end{pmatrix}$, $\mathbf{q}=\begin{pmatrix}5\\4\\6\end{pmatrix}$ and $\mathbf{r}=\begin{pmatrix}3\\-3\\-5\end{pmatrix}$.

Express in component form: (a) $\mathbf{p}+\mathbf{q}$, (b) $\mathbf{p}+\mathbf{r}$, (c) $\mathbf{q}+\mathbf{r}$, (d) $\mathbf{p}+\mathbf{q}+\mathbf{r}$, (e) $\mathbf{p}-\mathbf{r}$, (f) $\frac{1}{2}(\mathbf{p}-\mathbf{q}-\mathbf{r})$.

14 Calculate the modulus of the vectors:

$$\begin{pmatrix}3\\6\\6\end{pmatrix}, \begin{pmatrix}1\\-4\\8\end{pmatrix}, \begin{pmatrix}-2\\6\\9\end{pmatrix}, \begin{pmatrix}-4\\-6\\12\end{pmatrix}, \begin{pmatrix}12\\12\\14\end{pmatrix}, \begin{pmatrix}-8\\-8\\14\end{pmatrix}.$$

Triangles of velocities and forces.

15 Find by drawing, the resultants of the following pairs of velocities. (a) 18 m/s at 035° and 16 m/s at 085°. (b) 120 km/h at 325° and 35 km/h at 005°. (c) 40 km/h at 215° and 10 km/h at 135°. (d) 60 m/s at 200° and 45 m/s at 115°.

16 Find by drawing the components in the directions North and East of the forces (a) 15 N at 050°, (b) 20 N at 070°, (c) 18 N at 120°, (d) 16 N at 330°, (e) 12 N at 210°.

17 Draw triangles of force and hence find the equilibriant of the following pairs of forces.
(a) 60 N at 035° and 40 N at 075°, (b) 80 N at 270° and 55 N at 170°, (c) 12 kgf at 330° and 15 kgf at 050°.

18 Force $\mathbf{F}_1$ is 25 N at 030°. Force $\mathbf{F}_2$ is 35 N at 100°. Force $\mathbf{F}_3$ is 20 N at 320°. Find, by scale drawing, the resultants of (a) $\mathbf{F}_1 + \mathbf{F}_2$, (b) $\mathbf{F}_2 + \mathbf{F}_3$, (c) $\mathbf{F}_3 + \mathbf{F}_1$, (d) $\mathbf{F}_1 + \mathbf{F}_2 + \mathbf{F}_3$.

19 Using the forces from question 18 draw diagrams to illustrate that $(\mathbf{F}_1 + \mathbf{F}_2) + \mathbf{F}_3 = (\mathbf{F}_2 + \mathbf{F}_3) + \mathbf{F}_1 = (\mathbf{F}_3 + \mathbf{F}_1) + \mathbf{F}_2$.

20 Find the track and groundspeed of a light aircraft that flies on a course of 065° at an airspeed of 150 knots in a wind of 20 knots from 320°.

21 Repeat question 20 for the following aircraft.

	Course	*Airspeed*	*Wind Velocity*
(a)	040°	130 k	24 k from 320°
(b)	110°	140 k	25 k from due West
(c)	260°	120 k	20 k from 280°.

22 Tanker T steams due North at 15 k. Find the velocity relative to T of (a) ship A steaming due East at 12 k, (b) ship B steaming 155° at 10 k, (c) ship C steaming 340° at 20 k.

23 A small aircraft flies at an airspeed of 130 k in a wind of velocity 22 k from due South. Find the course to steer to make good a track of (a) 080°, (b) 035°, (c) 165°, (d) 265°.
State the ground speed in each case.

24 A sailor walks across the deck of an aircraft carrier at

6 km/h at right angles to the direction in which the ship is moving. If the carrier is sailing due South at 36 km/h, find the sailor's velocity relative to the ocean bed.

25 A man swims at 3 km/h, heading directly across a river which is flowing at 2 km/h. Find his velocity relative to the bank of the river.

How far is he carried downstream if the river is 60 m wide?

Position Vectors.

26 Using squared paper, mark the points with position vectors $\mathbf{v} = \begin{pmatrix} 3 \\ 1 \end{pmatrix}$ and $\mathbf{w} = \begin{pmatrix} 1 \\ 2 \end{pmatrix}$.

Indicate the points with position vectors (a) $2\mathbf{v}$, (b) $3\mathbf{w}$, (c) $-\mathbf{v}$, (d) $\mathbf{v} + \mathbf{w}$, (e) $\mathbf{v} - \mathbf{w}$, (f) $\frac{1}{2}(\mathbf{v} + \mathbf{w})$, (g) $\frac{1}{3}(\mathbf{v} + 2\mathbf{w})$.

27 $\mathbf{p} = \begin{pmatrix} 5 \\ 6 \end{pmatrix}$ and $\mathbf{q} = \begin{pmatrix} 3 \\ -2 \end{pmatrix}$ are the position vectors of points P and Q.

Find the co-ordinates of the points with position vectors (a) $2\mathbf{p}$, (b) $-\mathbf{q}$, (c) $\mathbf{p} + \mathbf{q}$, (d) $3\mathbf{p} + 4\mathbf{q}$, (e) $-\mathbf{p} - 2\mathbf{q}$, (f) $\frac{1}{2}(\mathbf{p} + \mathbf{q})$.

28 O is the origin and $\mathbf{OP} = \begin{pmatrix} 4 \\ 6 \end{pmatrix}$ and $\mathbf{OQ} = \begin{pmatrix} 6 \\ 2 \end{pmatrix}$.

Find the position vectors of the mid-points of (a) OP, (b) OQ, (c) PQ, (d) the mid-point of the line joining the mid-points of OP and OQ.

29 The position vector of a point P is $\begin{pmatrix} 6 \\ 2 \end{pmatrix}$. Find the position vector of the image of P after (a) a rotation of 90° anticlockwise, (b) a reflexion in the x-axis, (c) a rotation of 180°, (d) a reflexion in the y-axis followed by a reflexion in the x-axis.

30 $\mathbf{p}$ and $\mathbf{q}$ are the position vectors of the points P and Q. Write down the position vector of R a point on the line PQ if (a) PR = RQ, (b) PR : RQ = 1 : 3, (c) PR : RQ = 3 : 1, (d) PR : RQ = 3 : 2.

31 Point R has a position vector $\begin{pmatrix} 2 \\ 6 \end{pmatrix}$ and point S $\begin{pmatrix} 8 \\ 4 \end{pmatrix}$.

Write down the position vector of a point T on the line RS if RT : TS equals (a) 1 : 1, (b) 2 : 1, (c) 1 : 2, (d) 2 : 3.

32 What can be said about the quadrilateral whose vertices have position vectors $\mathbf{p}$, $\mathbf{q}$, $\mathbf{r}$ and $\mathbf{s}$ if (a) $\mathbf{p} - \mathbf{q} = \mathbf{s} - \mathbf{r}$, (b) $\mathbf{p} - \mathbf{s} = 2(\mathbf{q} - \mathbf{r})$?

33 $\mathbf{r} = \begin{pmatrix} 4 \\ 1 \\ 5 \end{pmatrix}$ and $\mathbf{s} = \begin{pmatrix} 3 \\ 2 \\ 4 \end{pmatrix}$ are the position vectors of points R and S. Find the co-ordinates of the points with position vectors (a) $2\mathbf{r}$, (b) $-\mathbf{s}$, (c) $\mathbf{r} + \mathbf{s}$, (d) $\mathbf{r} - \mathbf{s}$, (e) $\frac{1}{2}(\mathbf{r} + \mathbf{s})$.

34 For the points R and S in question 33, find the position vectors of the mid-points of (a) OR, (b) OS, (c) RS, where O is the origin.

Vector Geometry

Exercise 21

1 Copy and complete the following statements.
(a) $\mathbf{AB} = \mathbf{CD} \Leftrightarrow AB = \ldots$ and $\ldots$ is parallel to $\ldots$
(b) $\mathbf{AB} = \mathbf{BC} \Leftrightarrow \ldots = BC$ and ABC is $\ldots\ldots\ldots$
(c) $k\mathbf{AB} = l\mathbf{CD} \Leftrightarrow$ AB is . . . to . . . and $AB{:}CD = \ldots$
(d) $k\mathbf{AB} = l\mathbf{CD}$ and AB is not parallel to CD $\Rightarrow \ldots = 0$ and $\ldots = 0$.

2 In Figure 21.1, $AX{:}XB = CY{:}YB = 2{:}1$.
$\mathbf{AB} = \mathbf{p}$ and $\mathbf{BC} = \mathbf{q}$.
Express in terms of $\mathbf{p}$ and $\mathbf{q}$, (a) **XB**, (b) **BY**, (c) **XY**, (d) **AC**.
What can you say geometrically about (1) XY and AC, (2) figure XYCA?

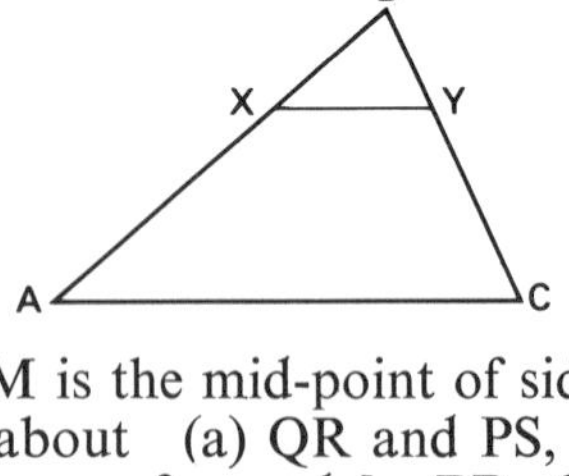

Fig. 21.1

3 In Figure 21.2 M is the mid-point of side PS. What can be said geometrically about (a) QR and PS, (b) figure PQRS?
(c) Express in terms of **a** and **b**: **PR**, **QS**, **RS** and **QM**.
What can be said about (1) QM and RS, (2) figure QRSM?

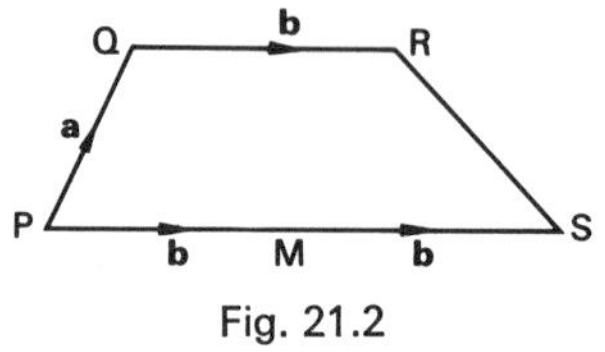

Fig. 21.2

4 ABCD, Fig. 21.3, is a quadrilateral with W, X, Y and Z the mid-points of sides AB, BC, CD and DA respectively. **AB** = **p**, **BC** = **q** and **CD** = **r**.

Express in terms of **p**, **q**, and **r** (a) **WB**, (b) **BX**, (c) **WX**, (d) **AD**, (e) **ZD**, (f) **ZY**.

What can be said about (1) WX and ZY, (2) figure WXYZ?

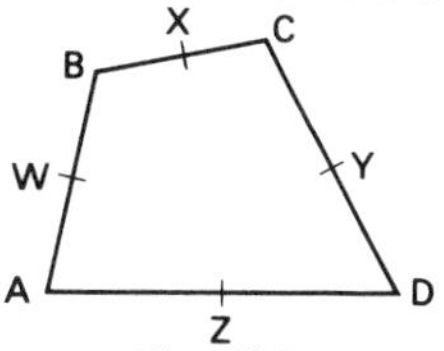

Fig. 21.3

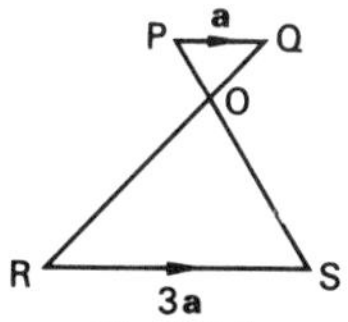

Fig. 21.4

5 Refer to Fig. 21.4. What can be said about (a) PQ and RS, (b) figure PQSR, (c) triangles POQ and SOR, (d) the ratio PO:OS?

6 In Figure 21.5 ABCD is a rectangle. M is the mid-point of BC and N a point on AD such that AN:ND = 1:3.

AB = **a** and **AD** = 4**b**.

Express in terms of **a** and **b**, **AN**, **BM**, **AM**, **BN** and **NM**.

What can be said about (a) BM and AN, (b) triangles AON and MOB, (c) the ratio BO:ON?

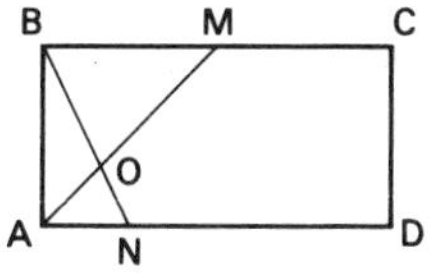

Fig. 21.5

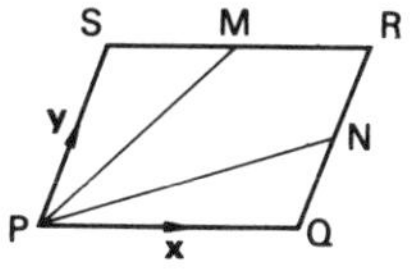

Fig. 21.6

7 PQRS, Fig. 21.6, is a parallelogram. M is the mid-point of RS and N the mid-point of QR. **PQ** = **x** and **PS** = **y**. Express in terms of **x** and **y**: **SR**, **RQ**, **PR**, **QS**, **SQ**, **PN**, **MP** and **MN**.

What is the relation between MN and SQ?

8 In Figure 21.7, RSTU is a parallelogram. RV = VW = WT. **RS** = **p**, **ST** = **q**. Express in terms of **p** and **q**, **RU**, **RT**, **RV**, **TW**, **VU** and **SW**. What can be said about (a) VU and SW, (b) figure VUWS?

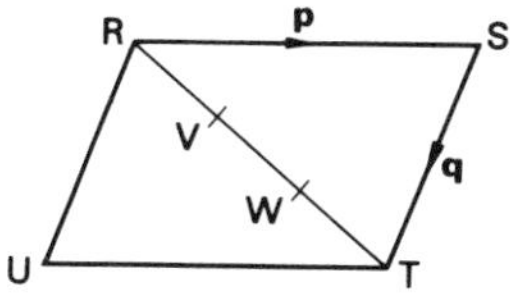

Fig. 21.7

9 In Figure 21.8, ABCD is a parallelogram. AB is produced to X so that AB = BX. CD is produced to Y so that CD = DY. **AB** = **p** and **AD** = **q**.

Express in terms of **p** and **q**: **AX**, **YC**, **AY** and **XC**.

What can be said about figure AYCX?

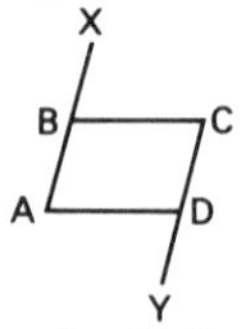

Fig. 21.8

10 PQRS, Fig. 21.9, is a quadrilateral with diagonals intersecting at O. **SO** = **OQ** = **a**, **RO** = **OP** = **b**. (a) Express in terms of **a** and **b**: **SP** and **RQ**. What can you say about SP and RQ?

(b) Express in terms of **a** and **b**: **PQ** and **SR**. What can you say about PQ and SR?

(c) What can you say about quadrilateral PQRS?

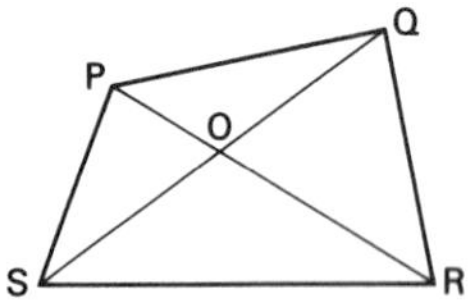

Fig. 21.9

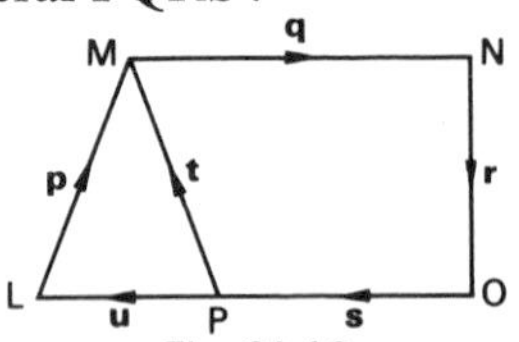

Fig. 21.10

11 Refer to Fig. 21.10. What can be said about (a) O, P and L if **s** = **u**, (b) OL and OP if **s** = **u**, (c) **s** + **q** if **r** + **t** = 0, (d) quadrilateral MNOP if **r** + **t** = 0, (e) the relation between **p**, **t** and **u**?

12 Use Fig. 21.11 for these questions.

(a) Express **ZW** in terms of **p** and **q**. (b) Express **ZW** in terms of **r**, **s** and **t**. (c) Give two reasons why **p** + **q** + **r** + **s** + **t** = 0.

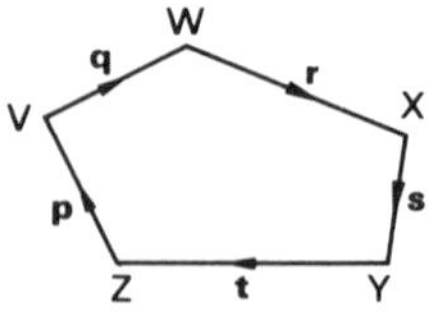

Fig. 21.11

13 Refer to Fig. 21.11 again.

(a) If **q** = **r**, what relations are there between the points V, W and X? (b) If **p** + **s** = 0, what can be said about (1) lines VZ and XY, (2) figure VXYZ? (c) If 2**r** + **t** = 0, what can be said about (1) lines WX and ZY, (2) figure WXYZ?

14 Use Fig. 21.12, for these questions.
(a) What can be said about (1) AB and DC, (2) figure BCDA? (b) Express in terms of **p** and **q**: **AC**, **AD**, **AE** and **BD**. (c) What can be said about AE and BD?

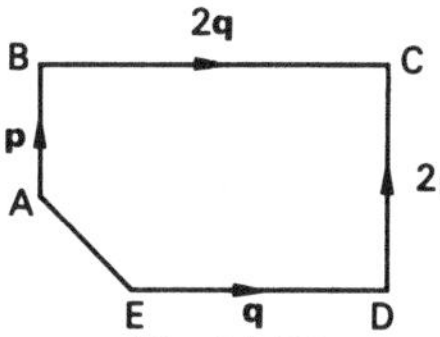

Fig. 21.12

15 ABCDEFGH is a cuboid. **AB** = **x**, **AD** = **y** and **AE** = **z**.
Express in terms of **x**, **y** and **z**: **BC**, **GF**, **DC**, **FE**, **AC**, **GE**, **HF**, **BD**.

16 For the cuboid in question 15 express in terms of **x**, **y** and **z** **AG**, **EC**, **HB** and **DF**.

17 PQRSTUVW is a cuboid. **PQ** = **p**, **PS** = **q** and **PT** = **r**.
Express in terms of **p**, **q** and **r**: **RS**, **RQ**, **WS**, **PR**, **WU** and **PV**.
Show that the line joining the mid-points of PQ and WV bisects the line joining the mid-points of TW and QR.

18 OABCD is a pyramid with a square base ABCD. **AD** = **p**, **AB** = **q** and **AO** = **r**. Express in terms of **p**, **q** and **r** (a) **BC**, **CD**, **AC** and **BD**, (b) **OA**, **OB**, **OC** and **OD**.

19 Refer to Fig. 21.13, for these questions.
(a) Express in terms of **p**, **q** and **r**: **AB**, **BC**, **DB**, **AV**, **CV** and **BV**.
(b) Do these results hold if VD is not at right angles to the base ABCD?

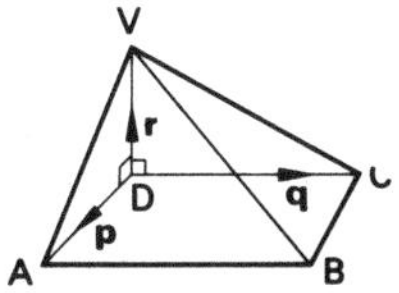

Fig. 21.13

20 In Fig. 21.13, mark M the mid-point of VB. Express in terms of **p**, **q** and **r**: **BM**, **DM**, **AM** and **CM**.

Answers

Exercise 1

1 (a) T (b) F (c) T (d) T (e) F (f) T (g) F.

3 (a) T (b) F (c) T (d) F (e) T (f) T (g) T (h) T.

4 33 26 8 40 25 64 14 23 37.

5 $(W \cup T)' \cap Y$ or $W' \cap T' \cap Y$
Plants taller than 1 m that flower in winter or have yellow flowers.

6 {King, Queen, Jack} {Cards numbered 2 3 . . . 8 9} {Ace, Ten}.

7 (a) V (b) I (c) I (d) I (e) V (f) I (g) V (h) I.

8 (a) T (b) F 74 (c) F 42 (d) T (e) F 123 (f) T.

9 $31 \neq 22 + 18 - 12$.
(a) 25 Saturday jobs (b) 21 Holiday jobs.

10 Yes $36 \neq 20 + 18 - 8$.

11 16. **12** 9. **13** 32. **14** 31 12.

15 (a) 4 (b) 3 (c) 4 26.

16 32 38 11 62 6.

17 21 24 51 51.

18 (a) 5 (b) 27 (c) 24 (d) 5 (e) 12 (f) 16.

19 (a) 101 (b) 100 111 000 (c) 3.

20 (a) Slide needle through right hand space. Cards that fall out are odd numbered.
(b) Slide needle through centre space and lift out cards. Slide needles through two outer spaces. Card that falls out is number 5. There are other methods.
(c) Slide needles through each of the three spaces. Card that falls out is number 7.

21 Needle through right hand space. Place cards that fall out, in order, at back of the pack. Repeat for middle and right hand spaces. Cards are now in ascending order.

22 (a) 8 (b) 16 (c) 64 (d) 256 (e) 1 024
(f) 1 048 576.

23 (a) 2 – 1 for each (b) Use 00 01 10 11 for the 4 houses
(c) 5.

24 (a) Needle through boy/girl space and lift out boys.
(b) Needle through boy/girl space and another through right hand house space. Pick up cards that fall out. Needle through left hand house space and lift out required cards.
(c) Needles through both house spaces. Pick up cards that fall out. Needle through lunch space of these cards. Lift out required cards.
(d) Needles through lunch space and mile space. Required cards fall out.

25

	M	T	W	Th	F	S	Sn
P	1	0	1	0	0	1	0
Q	1	0	1	0	1	1	1
R	0	1	1	1	1	1	0
S	1	0	1	0	0	1	1
T	1	1	1	1	1	1	1

(a) Wed Sat (b) Tues (c) Peter (d) Yes R
(e) (1) R (2) Q or R.

26

	E	F	I	R	G	S
A	1	1	1	0	0	0
B	1	0	0	1	1	0
C	1	1	0	0	0	1
D	1	1	0	0	1	1

(a) English (b) A C and D (c) E F G S
(d) A and B (e) C or D but preferable to keep C – hence D.

28 a b d e h.

29 (a) $\mathscr{E}$ (b) ϕ (c) PUQ (d) P (e) P (f) $\mathscr{E}$.

30 (a) ϕ (b) $\mathscr{E}$ (c) RUS (d) ϕ (e) P.

31 (Expression) $\cup$ (Dual) $= \mathscr{E}$
(a) $P' \cap Q'$ (b) $P \cap Q$ (c) $P \cup Q'$.

32 $P' \cap Q' \cap R$.

33 (a) $(P' \cap Q) \cup R$ (b) $P' \cap (Q' \cup R')$.

34 {2 6 8} {9 16} {4 16} {3 6} {2 6 8} {2 8}.

35 (b) $R - S = R \cap S'$ (d) $S - R = R' \cap S$
(e) It is ambiguous.

36 {2 6 8 9 16} {2 6 8 9 16} {2 3 6 8 9 16} {2 3 6 8 9 16}.

Exercise 2

1 (a) 337 (b) 33·7 (c) 1 880 (d) 1·81 (e) 0·397
(f) 2·52 (g) 0·396 (h) 39·6 (i) 774 (j) 43 300
(k) 3 760 (l) 2 400 (m) 4·29 (n) 13·56 (o) 0·429
(p) 24·9 (q) 10·8 (r) 1·11 (s) 3·84 (t) 0·828
(u) 0·543 (v) 0·0543 (w) 1·19 (x) 0·363 (y) 36·3
(z) 2·30.

2 (a) 228 cm² (b) 2 990 cm² (c) 1 150 cm²

3 109 52·3 245 1 635 yd
110 1 610 404 1 835 m.

4 56·6 88·9 117 1 410 lb
55·4 171 248 1 110 kg.

5 (a) 488 (b) 126 (c) 126 (d) 0·118 (e) 1·71
(f) 2 540.

6 (a) 31·4 243 24 300 0·314
(b) 841 2 810 11 200 112 424 35 000 cm²
(c) 91·6 125 211 21 100 973 97 300 cm².

7 (a) 1 730 12 200 12·2 91·1 91 100 2 250 000
(b) 2 200 4 910 22 000 50 700 111 636
1 910 cm³.
(c) 230 408 1 630 8 990 70 300 cm³.

8 (a) 13·3 42·1 15·2 4·82 8·33 2·63
(b) 4·24 6·48 8·66 27·4 34·6 60·6 cm.

9 (a) 3·85 3·87 5·13 11·1 2·61 12·1
(b) 5·50 11·7 2·52 5·44 cm.

10 97·4 151 176 490 4·90 cm.

11 726 940 2 460 6 380 cm².

12 42 93 76 60 33 95 51 89%.

13 61 76 87·5 21 51 96 82 67%.

14 104 108 109 110 112 km/h.

15 (a) 66 666 (b) 41 976 (c) 4 444 (d) 1 782
(e) 4 143 403 (f) 2·338 842 (g) 14 641
(h) 97 970 404 (i) 404·040 4.

16 25 627 10 423 69 279 67 319.

17 200·21 30·43 485·31 176·18.

18 18 445 14 161 24 025 2 194 955 0·767 741 9
1·302 521.

19 2 026·848 3 471·566 4 69 723·571 4 108 112·8
1·712 790 6 0·583 842 4.

20 1 022 390 152 185 492 37 604 22 073 548
344 569 114 017 1·857 594 9 2·655 462 1
0·202 725 7.

21 80·95 1 325·94 60·27 3 248·553 2 905·21
32 618·124 1 758 116·8 2·191 056 9 10·040 816
0·045 454 5.

22 (a) 1 052 m (b) 2 345 m (c) 3 996 m (d) 5 264 m.

23 (a) £796.35 (b) £291.20 (c) £196.24 (d) £419.97.

24 £74.80 £81.60 £93.50 £106.25.

25 (a) £72.60 £79.20 £90.75 £103.12½
(b) £68.20 £74.40 £85.25 £96.87½.

26 £2.70 £1.79 £0.67 £9.42 £352.10.

27 324 663 1 177 2 509 3 393 km/h.

28 \$239.59 \$452.01 \$587.86 \$1 768.52
\$3 766.25 \$268 244.47.

29 P 1 277 P 2 408 P 3 132 P 9 423 P 20 069
P 1 429 189.

31 Average speed = 12 132/s km/h 119 123 130
142 km/h.

32 Compute x (x − 3) x(x − 3) x(x − 3) + 2
240 −0·25 207·11.

33 (a) Compute x (x + 4) x(x + 4) x(x + 4) − 6
666 122·25 414·36.
(b) Compute x 2x (2x − 1) x(2x − 1)
x(2x − 1) + 3 123 1 543 15·88.

34 (a) 89 423 (b) 7 756 (c) 22 356 (d) £8 865.84
£6 228.80 £8 005.30 £7 303.88.
Add the 4 answers to part (d) and compare with the takings obtained from part (a).

35 (a) 29 100 km (b) 1 513 200 km (c) 13 200 km
(d) 843 km/h 924 km/h (e) 1 716 h.

36 (a) 749 kg (b) 4 622 kg (c) 1 659 kg (d) 47·4 kg
The smallest.

37 (a) Multiply by 0·405 (b) 3·40 1·30 2·63 2·31
1·94 ha (c) 11·58 ha (e) 2·47 acres.

38 (a) 255 379 452 420 570 808 549
(b) 3 433 (d) £299·38$\frac{1}{2}$.

39 (a) 89 836 (b) 6 370 (c) 24·5%.

40 (a) £220.48 (b) £107.81 (c) £261.67 (d) £162.46.

41 £5 725 £7 750 £10 000.

42 £300. **43** (a) £530 (b) £6 950. **44** 4·03 m.

Exercise 3

1 16 27 15 625 7 776 10 000 000 1 728.

2 2^5.

3 (a) 4^3 3^4 5^3 3^5 (b) 4^5 3^4 2^6 6^2.

4 0 5 9 9 6 3 6.

5 8 81 16 25 6 81 81 1 024.

6 43 $\frac{1}{18}$ 512 $\frac{16}{7}$.

7 a^5 b c^9 $2d^6$ $4e^6$ 1 g^3 h^3.

8 a^3b^3 b^2c^2 c^2d^4 1 $\frac{ac}{b}$ $\frac{a^3c}{b}$ $\frac{a}{bc^3}$.

9 $2^3 \times 5^6$ $2^{12} \times 5^{24}$ $2^{18} \times 5^{36}$ $2^2 \times 5^4$.

10 (a) 7 (b) 4 (c) 3.

11 (a) 6 $\frac{1}{6}$ 216 (b) 5 $\frac{1}{5}$ 25 (c) 4 2 $\frac{1}{4}$ $\frac{1}{16}$
(d) 9 3 27 $\frac{1}{9}$ 729 (e) $\frac{7}{8}$ $\frac{8}{7}$ $\frac{3}{4}$ $\frac{16}{9}$
(f) 0·1 0·2 5 0·2 25.

12 7 10 $\frac{1}{3}$ 4 343 $\frac{1}{2}$ $\frac{1}{216}$ $\frac{1}{2}$ 32 0·1 $2\frac{1}{2}$.

13 5 $2\frac{1}{2}$ $\frac{5}{3}$ 3 −3 $\frac{3}{2}$ 3 −3 −2.

14 4 2 4 $1\frac{1}{2}$ 4 3 4. **15** 4 1.

16 1 or 9 4 8 or 1.

17 2·492 0·623 $\bar{2}$·754 or −1·246 2·246 3·246.

18 (a) 4·96 1·24 $\bar{3}$·52 3·48 4·48
(b) 0·724 0·181 $\bar{1}$·638 1·362 2·362
(c) $\bar{1}$·28 $\bar{1}$·82 0·36 0·64 1·64.

19 3p $\frac{1}{2}$p −p p + 1 2p − 2.

20 2s −t s + t s + 2t 3s s + 4t s − t 2s − t.

21 2 m $\frac{1}{2}$ m 2 n −n m + n 2 m + n $\frac{1}{2}$ m − 2.

22 3 4 $\frac{1}{4}$ $\frac{2}{3}$ 2 3 −3 $\frac{1}{2}$ $\frac{3}{2}$ 2 3 −3 −2.

23 5 log x 2 log x 3 log x 9 log x.

24 3 2 0 3 0.

25 25 81 32 32 3 2 $\frac{1}{2}$.

26 20 20 5 1 or 2 $\frac{1}{25}$.

Exercise 4

1 (a) $12x + 8$ (b) $6x - 15$ (c) $4x^2 - 12x$
(d) $3x - 6x^2$ (e) $3x + 7$ (f) $2x^2 + 22x$
(g) $6x^2 + 11x - 10$ (h) $4x^2 + 11x - 3$
(i) $x^2 - 2x + 1$ (j) $9x^2 - 12x + 4$ (k) $4x^2 - 25$
(l) $6x - 3$.

2 (a) $p(p + 1)$ $3q(q + 5)$ $4r^2(1 - 9r)$ $\pi(R^2 + r^2)h$ $\frac{1}{2}h(s + t)$
(b) $(p + 1)(p + q)$ $(r - s)(r + 3)$ $(c + d)(e + 2f)$ $(a + b)(c + 2d)$
(c) $(x + 4)(x + 5)$ $(x - 3)(x - 6)$ $(4 - x)(2 + x)$ $(5 - x)(3 + x)$ $(2x + 1)(x + 3)$ $(3x - 2)(x + 4)$
(d) $(r + 6)(r - 6)$ $(s + 12)(s - 12)$ $(3t + 7)(3t - 7)$ $3(p + 5)(p - 5)$ $2(8 + q)(8 - q)$.

3 (a) 366 (b) 1 054 (c) 27 (d) 73 600 (e) 1 680
(f) 132 (g) 22.

4 (i) a^3b^3 $2b^6$ $8b^6$ $3c^2$ $\dfrac{2d}{c}$ $\dfrac{d^2}{e^2}$ fg^2

(ii) $\dfrac{7}{12a}$ $\dfrac{1}{12b}$ $\dfrac{(4+3c)}{12c^2}$ $\dfrac{(2-3d)}{6d^2}$

(iii) $\dfrac{(p+2)}{(p+1)(2p+3)}$ $\dfrac{-q}{(q+2)(q+4)}$ $\dfrac{r+7}{(r+2)(r+3)}$
$\dfrac{9}{(s-4)(s+5)}$.

5 (a) $\dfrac{4}{x(x-3)}$ (b) $2x(x-1)$ (c) $\dfrac{1}{x^2(x+1)}$

(d) $\dfrac{1}{x+1}$.

6 2 $\dfrac{x}{x-1}$.

7 (a) 1 $x-1$ (b) 1 $\dfrac{2}{x}$ (c) $\frac{1}{3}$ $\dfrac{x}{x+4}$ (d) $\frac{1}{3}$ $\dfrac{1}{x+1}$

(e) $1\frac{1}{7}$ $\dfrac{4x}{(2x+3)(2x-3)}$ (f) 1 $\dfrac{1}{x-1}$.

8 -3 9 24 12 36 13 1 1 5 25 5
$1\frac{1}{2}$ $-\frac{1}{6}$ 1 $-\frac{1}{6}$.

9 (a) $\dfrac{1}{x^2}$ $\dfrac{1}{2x}$ $2x^2$ $(2x)^2$ (b) $2x^2$

$(2x)^2 = \left(\dfrac{1}{2x}\right)$ $\dfrac{1}{x^2}$ (c) $\dfrac{1}{2x}$ $\dfrac{1}{x^2}$ $2x^2$ $(2x)^2$.

10 $\dfrac{mv^2}{F}$ $\dfrac{Fr}{v^2}$ $\sqrt{\dfrac{Fr}{m}}$ $\dfrac{V^2}{2h}$ $\dfrac{V^2}{2g}$ Id^2 $\sqrt{\dfrac{C}{I}}$ $\dfrac{p}{r+s}$

$\dfrac{p}{q}-s$ $\dfrac{p}{q}-r$ $\dfrac{V}{\pi r^2}$ $\pi r^2 h$ $\dfrac{mHt^2}{4\pi^2}$ $\dfrac{4\pi^2 I}{Ht^2}$ $\dfrac{4\pi^2 I}{mt^2}$.

11 (a) $\dfrac{3}{2-q}$ (b) $\dfrac{5}{4q-3}$ (c) $\dfrac{2(q+2)}{5-q}$

(d) $\pm\sqrt{\dfrac{q+2}{q-3}}$ (e) $\pm 2\sqrt{\dfrac{1-q}{1+q}}$.

12 (a) -2 $6\frac{3}{4}$ 28 (b) 2 or 5 1 or 6.

13 (a) -5 15 $-5\frac{5}{8}$ (b) $-\frac{1}{2}$ or 3 $3\frac{1}{2}$ or -1.

14 2 4 5 17 -15 3 6 11.

15 a a a c a b
$x = 6$ $y = -6$ $r = 6$ $s = 7$ $x = 3$ $y = 4$
$x = 4$ $y = 1$.

16 (a) 0 or -5 (b) -2 or 3 (c) $-\frac{1}{2}$ or $\frac{2}{3}$
(d) -2 -1 or 3.

17 (a) $\{-3\ 2\}$ (b) $\{-5\ -2\}$ (c) $\{-\frac{1}{3}\ 4\}$
(d) $\{-2\frac{1}{2}\ \frac{2}{3}\}$ (e) $\{-\frac{1}{5}\ 3\}$ (f) $\{-5\ \frac{3}{4}\}$
(g) $\{-1\frac{1}{4}\ 3\}$.

18 (a) 0 3 (b) 1 (c) any value (d) -5 1.

19 (a) 4·79 or 0·21 (b) $-6{\cdot}54$ or $-0{\cdot}46$
(c) $-4{\cdot}39$ or $-0{\cdot}11$ (d) $-1{\cdot}27$ or 2·77
(e) $-1{\cdot}13$ or 0·53 (f) $-0{\cdot}90$ or 0·74.

20 (a) 1·78 or $-0{\cdot}281$ (b) 1·27 or $-0{\cdot}472$
(c) $-1{\cdot}85$ or 0·181 (d) $-0{\cdot}906$ or 1·66.

21 (a) $x^2 - x - 1 = 0$ (b) $y^2 - 3y - 3 = 0$
(c) $2s^2 + 2s - 3 = 0$ (d) $4t^2 - 5t - 3 = 0$.

22 (a) $8x - 2$ (b) $2x + 13$ (c) $12x^2 - 24x - 15$
(d) $4x^2 - 20x + 25$ (e) $-6\frac{1}{2}$ (f) 1 or $5\frac{1}{2}$.

23 (a) 22 (b) $5x + 1$ (c) $\frac{L-5}{3}$ (d) $\frac{2(11-L)}{3}$
(e) 1 (f) 1·81 or $-$ 1·46.

24 (a) 27 18 15 $\frac{1}{3}$ 2 $4\frac{1}{3}$ 48 -24 -13
(b) $1\frac{1}{2}$ 0 2 $-\frac{2}{3}$ 2
(c) $\frac{R-3}{4}$ (d) $\frac{Q^2}{12}$ (e) $\frac{x+2}{6x^2}$ (f) $\frac{4x+1}{2x(4x+3)}$

25 (a) $x(x + 4)$ $(x - 1)(x + 4)$ $(x - 1)(x - 2)$
(b) 0 -4 (c) 1 (d) 1 or 8
(e) $x(x - 2)$
(f) $\frac{1}{x(x+4)(1-x)}$.

26 (a) 1 (b) -1 (c) $2\frac{1}{3}$ (d) -2 (e) $2\frac{7}{8}$
(a) -1 or 5 (b) 1 or 2 (c) 0 or $\frac{1}{4}$.

27 (a) 10 12 . . . $2n$ (b) 17 21 . . . $(4n - 3)$
(c) 5·7 6·8 . . . $n(n + 2)$
(d) $\frac{5}{6}$ $\frac{6}{7}$ . . . $\frac{n}{n+1}$ (e) $6x^5$ $7x^6$ . . . $(n + 1)x^n$
(f) 1 -1 . . . $(-1)^{n+1}$.

28 (a) 1 (b) all (c) 1 (d) 2 (e) all (f) 1.

29 (a) − (b) × (c) ÷ (d) +.

30 (a) × (b) + (c) ÷ (d) −.

31 (a) 65° 85° 95° 115° (b) 48° 96° 144° 72°
(c) 77° 77° 77° 129° (d) 75° 75° 95° 115°

32 (a) £$(x + 3)$ (b) £$(x - 2)$ £$(x + 5)$ (c) 9.

33 (a) 8 7 (b) 4 6.

34 (a) 9 (b) 64 (c) 25 (d) 36 (e) 36.

35 (a) The $(2x + 1)$ side (b) $5x$ $\frac{1}{2}x(2x - 1)$
(c) 8 units.

36 (a) 7 cm (b) 15 cm (c) 6 units (d) $3\frac{1}{2}$ cm.

37 (a) 420 500 −625 m (b) 5 or 15 8 or 12
0 or 20 s (c) 14·5 or 5·5 s.

38 1·62 : 1.

Exercise 5

1 15 15 9 25 DN DD ND NN NN
DD NN $2^{15} - 1$.

2

	−	+	×	0
3	(3−)	(3+)	(3×)	(30)
6	(6−)	(6+)	(6×)	(60)
9	(9−)	(9+)	(9×)	(90)

3

	2	4	6	8
1	(1 2)	(1 4)	(1 6)	(1 8)
3	(3 2)	(3 4)	(3 6)	(3 8)
5	(5 2)	(5 4)	(5 6)	(5 8)

(a) (1 2) (3 4) (5 6) (b) (3 6) (c) (1 4) (1 6)
(1 8) (3 6) (3 8) (5 8) (d) No pairs
(e) (3 4) (5 6).

4 30 60 72 72 25 144
(a) $n(Q) = n(R)$ (b) Q = R.

5

	2	4	12	24
1	(1 2)	(1 4)	(1 12)	(1 24)
6	(6 2)	(6 4)	(6 12)	(6 24)
12	(12 2)	(12 4)	(12 12)	(12 24)

(a) 1 (b) 3 (c) 4 (d) 4 (e) 5
(a) one-one (b) one-one (c) many-many
(d) many-many (e) many-many.

6 (a) $f: x \to \frac{1}{4}x$ (b) $f: x \to \frac{1}{2}(x - 1)$ (c) $f: x \to \sqrt[3]{x}$
(d) $f: x \to 5x - 3$.

9 10 −8 0 3 2 24 −96 8 $\frac{1}{6}$ $\frac{3}{4}$ 8·2
$\frac{12}{3x + 1}$ $\frac{36}{x} + 1$.

10 2, 5, −4, $\frac{7}{8}$, $\frac{1}{8}$; $\frac{1}{2} - \frac{1}{2}$, 8, $\frac{7}{8}$, $\frac{4}{7}$.
$\frac{x + 3}{2x}$, $\frac{8}{3x + 2}$, x, $\frac{9x + 14}{16}$

11 11 72 72 0 8 $\frac{1}{3}$ 50 17 2 $\frac{1}{8}$ 25
2·4 $2(2x + 1)^2$ $4x^2 + 1$ $\frac{32}{x^2}$ x $\frac{64}{x^2} + 1$
$\frac{2}{(2x + 1)^2}$.

12 R^+ and O Evens Multiples of 30 Odd integers
Square numbers doubled Triangular numbers
Reals from 0 to 36.

13 0 1 4 9 1 4 9.

14 $\frac{1}{2}$ 1 2 −2 −1 $-\frac{1}{2}$ $\frac{2}{0}$ is not defined.

15 15 5 −1 −3 −1 5 15.

17 4 2 1 $\frac{1}{2}$ $\frac{1}{4}$.

19 Reflexive because points (x x) lie on the graph.
Symmetric because the graph is symmetrical.
No, not transitive.

20 Reflexive and Symmetric yes.

21 (a) R S T (b) R S T (c) S (d) T
(e) R S T (f) S (g) T.

22 a b e. g, d – if the triangles are similar.

23 (a) T (b) S (c) R S T (d) T
d only c only.

24 (a) R S T (b) R S T – assuming one cycle per 6th former (c) T (d) S (e) S.

25 (1) a b (2) c (3) a b.

26 Ordering a b d e Equivalence c.

Exercise 6

1 y varies directly as x $y = Kx$ y varies inversely as x $y = \frac{K}{x}$

y varies inversely as the square of x $y = \frac{K}{x^2}$

y varies directly as the cube of x $y = Kx^3$

y varies directly as the square root of x $y = K\sqrt{x}$.

3 (a) y varies directly as the square of x (c) direct (d) increases 4 times (e) increases 44% (f) $y = \frac{2}{3}x^2$ 96.

4 (i) (a) y varies inversely as x (c) inverse (d) y is halved (e) decrease of $16\frac{2}{3}\%$ (f) $y = \frac{18}{x}$ $1\frac{1}{2}$

(ii) (a) y varies directly as the cube of x (c) direct (d) increase 8 times (e) increase of 72·8%

(f) $y = \frac{2x}{9}$ $2\frac{2}{3}$.

5 (a) $y \propto \sqrt{x}$ (c) increase 9 times (d) $y = 2\sqrt{x}$ (e) 18 1 296.

6 (a) $t \propto \sqrt{l}$ (b) take $\frac{1}{4}$ of the length (c) $t = 2\sqrt{l}$ (d) 3·6 s 4 m (e) increase of 20%.

7 (a) $s \propto t^2$ (b) $s = 5t^2$ (c) 125 m 6 s (d) $t = \sqrt{\frac{s}{5}}$

8 (a) $A \propto r^2$ (b) $C \propto r$ (c) $V \propto l^3$ (d) $N \propto \frac{1}{d}$ $N \propto \frac{1}{d^2}$

9 (a) $T \propto \frac{1}{v}$ (b) $N \propto \frac{1}{t}$ (c) $L \propto \sqrt{A}$ (d) $S \propto \frac{1}{l^2}$

10 $I \propto \frac{1}{d^2}$ (a) I decreases (c) Increases 4 times (d) decrease of 56% approx.

11 (a) $N \propto h^3$ (b) 80 000 33 750 156 250.

Exercise 7

1 (a) $<$ (b) $>$ (c) $<$ (d) $>$ (e) $=$ (f) $<$
(g) $<$ (h) $>$ (i) $< <$ (j) $< <$.

2 (a) $>$ (b) $>$ (c) $>$ (d) $<$ (e) $>$
(f) no conclusion (g) no conclusion (h) $<$.

3 (a) $\Leftrightarrow$ $<$ (b) $<$ $\Leftrightarrow$ (c) $\Leftrightarrow$
(d) $\Leftrightarrow$ $<$ (e) $\Rightarrow$ $<$ (f) $\Leftrightarrow$ $<$
(g) $\Rightarrow$ $>$.

4 (a) $n > q$ (b) $p \geqslant 7$ (c) $p < 5$ (d) $n \leqslant p$
(e) $q > o$.

5 (a) $x > 2$ (b) $x \leqslant 4$ (c) $x < 5$ (d) $x \leqslant 3$
(e) $x \geqslant -4$ (f) $x < 27$.

6 (a) $y \leqslant 1{\cdot}8$ (b) $y \geqslant 5\frac{1}{2}$ (c) $y < 6\frac{1}{2}$ (d) $y \leqslant 5\frac{1}{3}$
(e) $y > 5\frac{2}{3}$ (f) $y \leqslant 21\frac{2}{3}$.

7 (a) $x \leqslant 7$ (b) $x = 6$ or 3 (c) $x = 6$.

8 (a) $x \leqslant 7$ (b) 7 6 5 4 3 2 1 0 (c) 6 4 2.

9 (a) $x \leqslant 17$ (b) 15 10 5
(c) 7 13 11 7 5 3 2.

10 (a) $x \geqslant 6$ (b) 6 10 15 30
(c) 6 8 12 16 24 48.

11 (a) $4 < x < 9\frac{1}{2}$ (b) 5 7 9 (c) 6 9.

12 (a) $5 \leqslant x \leqslant 23$ (b) 5 7 11 13 17 19 23
(c) 6 12 18.

13 (a) $-6 < y < 4$ (b) $-4 < y < 3$ (c) $-4\frac{1}{2} \leqslant y \leqslant \frac{1}{2}$
(d) $-3 \leqslant y \leqslant 3$ (e) $-13 < y < 23$.

14 (a) Any negative value of x (b) $x \geqslant \frac{1}{4}$ or $x < 0$
(c) $x > 4$ $x < 0$ $x > 1\frac{1}{2}$ $x < 0$ $x > \frac{3}{5}$ $x < 0$
$0 < x < \frac{3}{5}$.

15 (a) $0 < x < 10{\cdot}8$ (b) $0 < x \leqslant \frac{1}{9}$
(c) $x \geqslant \frac{5}{13}$ a and b: no further solutions;
c: $x < 0$ is also a solution.

18 $32 \leqslant n \leqslant 48$.

19 $56 \leqslant v \leqslant 72$ km/h.

20 Length $< 5{\cdot}5$ m Width $< 4{\cdot}5$ m Weight < 10 tonne.

21 (a) $68 \leq$ perimeter ≤ 72 cm
(b) $282\frac{3}{4} \leq$ area $\leq 317\frac{3}{4}$ cm^2.

22 $8 \leq v \leq 10$ $90 \leq n \leq 140$.

Exercise 8

1 (a) 3·6 cm (b) 5 cm (c) 2·1 h (d) 50 km/h

2 (a) 8·6 5·0 2·7 (b) 6·7 m/s (c) −5·3 (d) 2.

3 −1 speed.

4 0·4
Draw $y = \frac{1}{x}$ for negative values and read off the value of x where this curve cuts $y = x + 2$.

5 (a) 11 22 35 m/s (b) 1·2 m/s^2 (c) 2 380 m.

6 (a) 10·5 8 6 1/min
(b) After $2\frac{1}{2}$ min (c) 143 1.

7 (a) 240 390 330 m/s (b) 32 s (c) 22 500 m.

8 $\frac{1}{9}$ 1 27 (a) 1·7 5·2 (b) 1·9 2·65 (c) 2·75.

9 (a) 3·8 5·8 22 (b) $10\frac{1}{2}$ days.

10 (a) 300 (b) 390 205 (c) 32 15 8 penguins per year
Decreasing.

11 (4 4) (4 8) (4 12) (8 4) (8 8)
(a) (8 8) (b) (4 12).

12 (3 3) (3 6) (3 9) (3 12) (6 3) (6 6) (6 9)
(6 12) (9 3) (9 6) (a) (6 12) or (9 6) (b) (6 12).

13 £27 £54. **14** £20 £60. **15** 34 28.

16 4 horses 6 ponies.

17 2 horses and 8 ponies 3 and 7 4 and 6.

18 (2 2) (2 4) (2 6) (2 8) (4 2) (4 4) (4 6)
(6 2) (6 4) (8 2)
(1) (8 2) (2) (2 8).

19 (a) (3 3) (3 5) (3 7) (5 3) (5 5) (7 3)
(1) (7 3) (2) (3 7)
(b) (3 3) (3 6) (6 3) (1) (6 3) (2) (3 6).

20 4 pairs of budgerigars   1 pair of canaries.

21 Use 8 of X and 2 of Y
(8 2) £42 (7 3) £43 (6 4) £44.

22 (a) 17 (b) 4 gnomes and 6 dwarfs 6 and 3
(c) 10 for (3 7) or (4 6) (d) 3 gnomes and 7 dwarfs.

Exercise 9

1 (a) 5 m (b) 6·4 m (c) 6·16 m (d) 5·04 m
(e) 5·28 m (f) 4·4 m.

2 (a) 100 cm^2 (b) 78·5 cm^2 (c) 314 cm^2 (d) 43·3 cm^2
(e) 260 cm^2.

3 (a) 100 m (b) 80 m (c) $83\frac{1}{3}$ m (d) 56·4 m
(e) 152 m.

4 (a) 140 cm^2 (b) 76 cm (c) 9·43 cm.

5 120 cm^2.

7 (a) 40 cm^2 (b) 40 cm^2
A parallelogram is equal in area to a rectangle with the same base and between the same parallels.

8 100π 400π 500π 400π cm^2.

9 48 cm^2. **10** (d) 90 cm^2 (e) 36 cm.

11 (a) 84 cm^2 (b) 35 cm^2 (c) 154 cm^2 (d) 13·9 cm^2.

12 (b) Triangles WXY and WZY WOX and WOZ
YOX and YOZ (c) 21 cm (d) 168 cm^2.

14 (b) 39·3 cm^2 (c) 31·4 cm (d) Also 31·4 cm.

15 (b) 66 cm^2 (c) £55.

18 10·4 cm 6.

Exercise 10

1 (a) 12 cm^2 (b) 5 cm (c) 106° 16′ (d) 53° 08′
126° 52′.

2 (a) 8 cm (b) 120 cm^2 (c) 123° 50′ (d) 61°55′
118° 05′.

3 (a) 120° 120° (b) 20·8 cm (c) 62·4 cm².

4 (a) 29 cm (b) 87° 12′ (c) 43° 36′ 136° 26′.

5 (a) 116° 58° (b) 76° 52° 19° (c) 24° 66° 33° (d) $22\frac{1}{2}$°.

6 (a) 60° 120° 60° (b) 18° 36° 54° (c) 52° 64° 38°.

7 (a) 150° 105° 75° (b) 58° 116° 122° (c) 70° 110° 70° (d) 34° 34° 56° 32°.

8 (a) 132° 66° 114° (b) 144° 108° 36° (c) 44° (d) 128° 116° 52° (e) 25° (f) $\frac{1}{2}x°$ $180° - \frac{1}{2}x°$ $180° - x°$.

9 (a) 50° 100° 90° (b) 128° 64° 90° (c) 60° 60° 60° (d) 56°.

10 (a) Triangles WOZ and XOY WOX and ZOY (b) 8 20 8·8 cm 5 or 12 cm.

11 (c) 6 cm 4·2 cm $2\frac{1}{4}$ cm 4.

12 (b) 12 $6\frac{2}{3}$ 13·4 4 cm.

13 10 18 cm. **14** 10 m. **15** 101 cm².

16 9 cm 1 cm 9 cm possible impractical.

17 4 or 100 cm.

18 (a) 6·7 cm (b) 5·3 cm (c) 7·75 cm (d) 5·9 cm.

Exercise 11

1 (a) 80 000 cm² (b) 1·584 m³ (c) 38·8 m³.

2 (a) 216 cm³ (b) 78 cm³ (c) 704 cm³ (d) 235 cm³ (e) 50 cm³ (f) 1 437 cm³.

4 280 140 157 120 1 540 513 524 408 cm³.

5 (a) 1 3 3 (b) 1 3 5 (c) 0 4 4.

6 (a) 1 1 ∞ (b) 0 1 ∞ (c) 1 ∞ ∞ (d) 1 13 9.

7 (a) 88 cm (b) 26 400 cm² (c) 77·6 kg (d) 74·4 cm.

8 (a) 10·5 cm (b) 10 200 cm³.

9 0·495 m^3 more.

10 (a) 693 m^3 (b) 132 m^2 (c) 479 m^2 (d) 58·2 m^3.

11 (a) The cylinder (b) Equal (c) 5 240 cm^3
(d) 1 570 cm^2.

12 2 2 3 2 3·61 cm.

13 4 cm 3 spheres 8 spheres.

14 No.

15 (a) $\frac{1}{3}$ (b) $\frac{\pi}{4}$ (c) $\frac{\pi}{6}$ (d) $\frac{2}{3}$.

16 (a) $\frac{1}{2}$ (b) $\frac{1}{3}$ (c) $\frac{32}{45}$.

17

3:1	1:12	5:3	1:8	1:5	7:2	1:10
9:1	1:144	25:9	1:64	1:25	49:4	1:100
27:1	1:1728	125:27	1:512	1:125	343:8	1:1 000

18 (a) 1:24 (b) 1:576 (c) 1:13 824 (d) 1:24
(e) 1:13 824.

19 (a) 5:6:9 (b) 25:36:81 (c) 125:216:729.

20 711. **21** $5\frac{1}{3}$.

22 (a) 1:400 (b) 1:20 (c) 1:8 000 (d) 1:400
(e) 1:8 000.

23 736 000 432 cm^2.

24 (a) 25:4 (b) 5:2 (c) 125:8 (d) 125:8 £106.25.

27 b and d only. **28** a and b only. **29** The cube.

30 Right angled, scalene; right angled, scalene; acute, isosceles; right angled, scalene; obtuse, scalene; right angled, scalene; obtuse, scalene; acute, isosceles.

31 AX AH AF AY AC AG.

32 24·1 cm. **33** 135°.

34 (a) 1 392 cm^3 (b) 87 cm^2 104 cm^2.

35 20° 33′ 16° 42′ 134° 46′ 39° 59′.

36 (b) $20\frac{1}{2}$° $10\frac{1}{2}$° (c) 80 cm (d) 42 000 cm^3.

37 (a) 7·5 m^2 (b) $2\frac{1}{2}$ m^2 (c) $7\frac{1}{2}$ m^3.

38 38° 40′. **39** (a) 1 (b) 3·66 m **41** 1·3 m.

42 448 56 112 112 112 56 cm³ $\frac{7}{8}$.

43 (b) $46\frac{1}{2}°$ 32° (c) 15·8 cm.

44 (b) 395 cm² (c) 20·2 cm.

45 (c) 13 cm. **46** 25:24. **47** 62·8 cm².

Exercise 12

1

sin	+	+	−	−
cos	+	−	−	+
tan	+	−	+	−

2 (a) all sine tan cos
(b) cos and tan cos sin and cos sin and tan
sin and tan.

3

sin	0·7660	0·3420	0·1736	0·8192	0·5736
cos	−0·6428	−0·9397	−0·9848	−0·5736	−0·8192
tan	−1·1918	−0·3640	−0·1763	−1·4281	−0·7002

sin	0·2588	0·8910	0·8090	0·3907	0·2250
cos	−0·9659	−0·4540	−0·5878	0·9205	0·9744
tan	−0·2679	−1·963	−1·3764	0·4245	0·2309

4

sin	0·9502	0·8355	0·5937	0·2638	0·1516
cos	−0·3118	−0·5495	−0·8047	−0·9646	−0·9885
tan	−3·047	−1·5204	−0·7378	−0·2736	−0·1533

5

sin	−0·3420	−0·9397	−0·9848	−0·7660
cos	−0·9397	−0·3420	0·1736	0·6428
tan	0·3640	2·747	−5·671	−1·1918

sin	−0·2588	−0·1045
cos	0·9659	0·9945
tan	−0·2679	−0·1051

6 30° 150° 210° 330° 45° 315° 150° 210°
45° 225° 135° 315°.

7 44° 26′ 135° 34′ 203° 35′ 336° 25′ 78° 28′
281° 32′ 107° 28′ 252° 32′ 58° 238° 107° 21′
287° 21′.

8 (a) sin $x°$ 0·00 0·50 0·71 0·87 1·00
cos $x°$ 1·00 0·87 0·71 0·50 0·00
(b) sin $x°$ 0·87 0·71 0·50 0·00
cos $x°$ –0·50 –0·71 –0·87 –1·00
(c) sin $x°$ –0·50 –0·71 –0·87 –1·00 0·87 –0·71
–0·50 0·00
cos $x°$ –0·87 –0·71 –0·50 0·00 0·50 0·71
0·87 1·00.

9 (a) 0·00 0·58 1·00 1·73 5·67
(b) –1·73 –1·00 –0·58 0·00
(c) 0·58 1·00 1·73 5·67 –1·73 –1·00 –0·58
0·00.

15 One is the reflection, in the x axis, of the other.

16 (a) graph rises 1 unit (b) graph rises 4 units
(c) lowered 2 units (d) lowered 4 units.

17 (a) $\sin(x° + 90°) = \cos x°$ (b) $\cos(x° + 90°) = -\sin x$.

18 (a) 2 times and 4 times the amplitude
(b) 3 times and $\frac{1}{2}$ times the amplitude
Stretched parallel to the y axis.

19 (a) Oscillates twice as rapidly and half as rapidly
(b) Oscillates 3 times as rapidly and $\frac{1}{3}$ as rapidly
Stretched parallel to the x axis.

20 (a) $y = -\sin x°$ (b) $y = 1 + \sin x°$ (c) $y = \cos x°$
(d) $y = \sin x°$ (e) $y = 1 - \sin x°$ (f) $y = \tan x°$
(g) $y = 1 + \cos x°$.

21 (a) $y = 2 \sin x°$ (b) $y = -\tan x°$
(c) $y = 1 + 2 \sin x°$.

23 (a) 1·28 1·16 0·12 0·12 (b) 1·41 –1·41
(c) $17\frac{1}{2}$ or $162\frac{1}{2}$ $107\frac{1}{2}$ or $252\frac{1}{2}$ 123 or 327.

24 (a) 2·24 –2·24 (b) 96 317 $95\frac{1}{2}$ $317\frac{1}{2}$.

25 (a) 1·41 –1·41 (b) 70 200 20 250.

26 50 and –50 units 2·5 14·5 s.

27 26 and –26 units 0·75 6·75 s.

28 38. **29** 63.

30 (b) 0·64 0·98 –0·98 –0·64
(c) 27 63 207 243 117 153 297 333
(d) 78.

Exercise 13

1 51·0 82·8 52·7 19·4 cm².

2 (a) 23·0 2 300 cm² . The areas are in the ratio 1 : 100
(b) 78·9 78·9 cm² . The areas are equal
(c) 143 143 cm² . Again equal
(d) 18·0 27·6 cm² . No relation.

3 (a) 55·2 cm² (b) 106 cm² (c) 124 cm².

4 46·2 cm². **5** (b) 72° (c) 47·6 238 cm².

6 (a) 260 cm² (b) 283 cm² (c) 294 cm².

7 1 040 1 130 1 180 cm².

8 (a) 9·80 cm² (b) 21·3 cm² (c) 87·8 cm²
(d) 8 780 m² (e) $2{\cdot}23 \times 10^6$ m².

9 187 ha. **10** 52·9 m².

11 (1) 19·1 (2) 13·7 (3) 22·0 units.

12 (1) 51° 33′ or 128° 27′ (2) 37° 48′ or 142° 12′.

13 (1) 15·7 cm (2) 11·2 cm (3) 15·5 cm (4) 24·7 cm.

14 104 cm. **15** (1) 11·5 cm (2) 10·1 cm (3) 160 m.

16 64° 09′ or 115° 51′.

17 (a) 9·92 cm (b) 40·4 cm.

18 (a) 3 050 m (b) 1 920 m.

19 (a) 110° 22° (b) 119 m (c) 112 m.

20 68·7 m. **21** (b) 50·9 m.

22 (b) 25° 55° 100° (c) P (d) 343 m.

23 (1) 8 cm (2) 9·64 cm (3) 8·72 cm (4) 7·55 cm.

24 (a) 8·08 cm (b) 14·9 cm (c) 26·5 cm (d) 44·7 cm.

25 (a) 82° 49′ (b) 78° 28′ (c) 110° 45′ (d) 104° 28′.

26 (a) Acute (b) Obtuse (c) Obtuse (d) Right-angled
(e) Acute.

27 21·8 cm 31·1 cm.

28 (a) 7·94 cm (b) 13·1 cm (c) 93·5 cm².

29 068° 12′. **30** 19·7 km.

31 26·3 52·6 65·7 m. **32** 2·18 m.

33 1·84 m. **34** BEN'S.

35 (a) $\frac{4}{5}$ $\frac{3}{4}$ (b) 21 cm^2 (c) 6·08 cm.

36 (a) $\frac{5}{13}$ $\frac{12}{5}$ (b) 48 cm^2 (c) 12·4 cm.
(a) $\frac{8}{17}$ $\frac{15}{8}$ (b) 75 cm^2 (c) 15·1 cm.

37 (a) 78° 28′ 57° 07′ 44° 25′ (b) Both as part (a)
(c) 032° 53′ 147° 07′.

38 (a) $\frac{1}{4}$ (b) 9·695 cm (c) 34·9 cm^2.

39 (a) 16·2 cm (b) 7 cm (c) 113 cm^2.

40 64·3 m.

Exercise 14

1 small great great small great small great small

2 (a) 30° (b) 50° (c) 18° (d) 9° (e) 61° (f) 138°.

3 (a) 12° (b) 21° (c) 56° (d) 52° (e) 102°.

4 5 510 4 500 3 180 2 690 1 430 km.

5 60° 75° 32′ 71° 40′ 61° 52′ 83° 30′.

6 6 660 2 664 4 995 7 992 8 880 1 776 km.

7 1 880 1 510 1 400 503 km.

8 3 330 5 000 1 110 5 550 km.

9 3 845 743 1 010 km.

10 19° 06′ 9° 33′ 5° 44′ 17° 11′.

11 19° 06′ 17° 11′ 34° 23′.

12 (a) 40 000 km (b) 28 300 km (c) 60° (d) 84° 16′.

13 (a) 3 110 km (b) 6 880 km (c) 13 100 km
(d) 6 220 km.

14 20·8 km/h. **15** 7·54 h 2·01 h.

16 1 110 km 668 km.

17 (a) 3 470 km (b) 2 220 km.

18 (a) 3 916 km (b) 683 km (c) 11° 54′ W.

19 (a) 24° 17′ W (b) 18° 14′ E.

20 (a) 95·1 cm (b) 126 cm (c) 59·8 cm.

21 (a) 2 730 km (b) 1 120 km (c) 1 410 km.

22 (a) 11 300 km (b) 7 140 km.

Exercise 15

1 (a) $\frac{1}{2}$ (b) $\frac{1}{4}$ (c) $\frac{3}{13}$ (d) $\frac{1}{13}$ (e) $\frac{1}{26}$ (f) $\frac{1}{52}$
(g) $\frac{3}{4}$.

2 (a) $\frac{1}{6}$ (b) $\frac{1}{2}$ (c) $\frac{1}{3}$ (d) $\frac{1}{2}$ (e) $\frac{5}{6}$ (f) $\frac{5}{6}$.

3 (a) $\frac{1}{6}$ (b) $\frac{1}{5}$ (c) $\frac{1}{6}$ (d) $\frac{1}{10}$.

4 (a) $\frac{1}{9}$ (b) $\frac{4}{9}$ (c) $\frac{5}{9}$ (d) $\frac{1}{3}$
The proprietor.

5 (a) $\frac{1}{600}$ (b) $\frac{1}{30}$ (c) $\frac{1}{24}$ (d) $\frac{3}{1000000}$.

6 $\frac{2}{15}$ $\frac{1}{100}$.

7 (a) equal (b) second (c) first (d) equal (e) first
(f) second (g) equal.

8 (a) $\frac{1}{26}$ (b) $\frac{5}{26}$ (c) $\frac{3}{13}$ (d) $\frac{7}{26}$.

9 (a) $\frac{3}{5}$ (b) $\frac{2}{5}$ (c) $\frac{1}{5}$ (d) $\frac{4}{5}$.

10 (a) $\frac{3}{10}$ (b) $\frac{1}{10}$ (c) $\frac{1}{10}$ (d) $\frac{7}{10}$.

11 (a) $\frac{1}{4}$ (b) $\frac{1}{8}$ (c) $\frac{3}{8}$.

12 (a) $\frac{1}{16}$ (b) $\frac{7}{16}$ (c) $\frac{3}{8}$.

13 (a) $\frac{1}{2}$ (b) $\frac{1}{4}$. **14** (a) $\frac{1}{6}$ (b) $\frac{1}{6}$.

15 (a) $\frac{16}{221}$ (b) $\frac{16}{221}$ (c) $\frac{1}{221}$ (d) $\frac{33}{221}$.

16 (a) $\frac{1}{6}$ (b) $\frac{1}{2}$ (c) $\frac{1}{12}$ (d) $\frac{7}{12}$.

17 (a) $\frac{1}{36}$ (b) $\frac{5}{36}$ (c) $\frac{1}{6}$ (d) $\frac{1}{4}$.
7 1.

18 (a) $\frac{1}{6}$ (b) $\frac{1}{2}$ (c) $\frac{1}{3}$ (d) $\frac{1}{4}$ (e) $\frac{1}{12}$ (f) $\frac{1}{6}$.

19 (a) Odd (b) Even (c) $\frac{1}{5}$ $\frac{1}{25}$ $\frac{1}{125}$
Very unfair.

20 $\frac{2}{3}$. **21** $\frac{4}{9}$. **22** $\frac{8}{15}$. **23** $\frac{1}{3}$ $\frac{1}{9}$ $\frac{1}{5}$.

24 1 (a) 0·01 (b) 0·16 (c) 0·04 (d) 0·25 No.

Exercise 16

5 213 000 1970 1971.

6 (a) 2 5 5 6 7 8 9 7 eggs
(b) 13 21 37 46 72 59
(c) 12 17 18 21 22 27 32 43 31 days
(d) 3 4 5 5 6 6 7 8 5 peas
(e) 110 000 120 000 130 000 140 000 150 000
40 000 bars.

7 (a) 6 eggs (b) 37 (c) $21\frac{1}{2}$ days (d) $5\frac{1}{2}$ peas
(e) 130 000 bars.

8 (a) 6 eggs (b) 37·8 (c) 24 days (d) $5\frac{1}{2}$ peas
(e) 130 000 bars.

9 (a) 12 12 (b) 13 11 (c) 12 14 (d) 2·2 2·25.

11 (a) 12·16 (b) 10·35 (c) 14·27 (d) 2·25.

12 (a) 2 4 8 10 6 3 2 1 (b) 19 12 cm
(c) 15 cm (d) 15 cm.

14 15 cm.

15 (a) $\frac{1}{36}$ (b) $\frac{1}{18}$ (c) $\frac{1}{6}$ (d) $\frac{1}{3}$ (e) $\frac{1}{1296}$.

16 (a) 1 4 7 11 8 4 3 2 (b) 105–109 s
(c) 105 s (d) $12\frac{1}{2}\%$.

17 Frequency polygon.

18 109 s.

19 (a) 18 060 : 23 999 miles (b) 19 250 miles.

20 24 080 miles.

21 (a) $\frac{4}{75}$ (b) $\frac{43}{75}$ (c) $\frac{4}{75}$.

22 (b) 15–19 times 39 times (c) 16%.

23 (a) 2 7 12 17 22 27 32 37 times (c) 18 times.

24 (a) 17 times (b) 6 times.

26 (a) 7 (b) 7 6.90. **27** 1·7.

29 (a) 4 (b) 4 4·2.

30 (a) $\frac{3}{20}$ (b) $\frac{1}{24}$ (c) $\frac{1}{10}$ (d) $\frac{59}{60}$.

31 3·7 occupants per house
1·2 occupants per house.

32 15 years 4·9 months.

33 69·6 cm. **34** £1 253.

Exercise 17

1 (a) 3×2 2×3 (b) 3×3 (c) $\begin{pmatrix} 3 & 7 & 11 \\ 6 & 14 & 22 \\ 9 & 21 & 33 \end{pmatrix}$

(d) Yes (e) No (f) Only 2**N**.

2 (a) 2×3 3×1 2×2 1×2
(b) 2×1 1×2 2×3
(c) $\begin{pmatrix} 36 \\ 78 \end{pmatrix}$ $(18 \quad 12)$ $\begin{pmatrix} 26 & 34 & 42 \\ 27 & 36 & 45 \end{pmatrix}$
(d) No no no yes (e) No (f) $\mathbf{Y}^2$.

3 **K** is a square matrix.

4 $\begin{pmatrix} 11 & 10 & 11 \\ 10 & 11 & 10 \end{pmatrix}$ Not possible not possible $\begin{pmatrix} 11 & 10 \\ 21 & 21 \\ 10 & 11 \end{pmatrix}$

$\begin{pmatrix} 5 & 4 & 5 \\ 9 & 9 & 9 \\ 4 & 5 & 4 \end{pmatrix}$ $\begin{pmatrix} 9 & 9 \\ 9 & 9 \end{pmatrix}$.

5 Not possible $\begin{pmatrix} -3 & 4 & 3 \\ -2 & 4 & 5 \end{pmatrix}$ not possible

$\begin{pmatrix} 18 & 19 & 38 \\ 4 & 7 & 4 \end{pmatrix}$ $\begin{pmatrix} -19 & 35 \\ -32 & 22 \end{pmatrix}$ $\begin{pmatrix} 4 & 12 \\ 8 & 16 \\ -24 & 8 \end{pmatrix}$

$\begin{pmatrix} 8 & 6 \\ 4 & -2 \end{pmatrix}$ $\begin{pmatrix} -3 & 12 & 11 \\ -2 & 9 & 10 \end{pmatrix}$ $\begin{pmatrix} 22 & 9 \\ 6 & 7 \end{pmatrix}$ Not possible

not possible $\begin{pmatrix} -759 & 105 \\ 96 & -636 \end{pmatrix}$.

6 Yes yes no no no yes no yes.

7 (a) $(14 \quad 10 \quad 11)$ (b) $\begin{pmatrix} 9 \\ 6 \\ 9 \\ 11 \end{pmatrix}$ (a) adds the columns
(b) adds the rows

(35) (35) Either method gives the sum of the elements in **M**.

8 $\begin{pmatrix}19 & 47\\36 & 79\end{pmatrix}$ $\begin{pmatrix}36 & 32\\66 & 55\end{pmatrix}$ $\begin{pmatrix}81 & 28\\142 & 48\end{pmatrix}$ $\begin{pmatrix}2 & -1\\-5 & 3\end{pmatrix}$.

9 (a) $\begin{pmatrix}15 & 32\\26 & 55\end{pmatrix}$ $\begin{pmatrix}63 & 12\\113 & 23\end{pmatrix}$ $\begin{pmatrix}10 & 65\\17 & 115\end{pmatrix}$ $\begin{pmatrix}53 & 36\\93 & 63\end{pmatrix}$

$\begin{pmatrix}27 & 3\\51 & 5\end{pmatrix}$ (b) DOPE.

10 (a) HUSH (b) MICE (c) HELP.

11 $\begin{pmatrix}3 & -2\\-7 & 5\end{pmatrix}$ $\begin{pmatrix}2 & -3\\-5 & 8\end{pmatrix}$ $\begin{pmatrix}-3 & -4\\-5 & 7\end{pmatrix}$ $\begin{pmatrix}2 & -5\\-1 & 3\end{pmatrix}$

$\begin{pmatrix}3 & -2\\-4 & 3\end{pmatrix}$ $\begin{pmatrix}4 & -5\\5 & -6\end{pmatrix}$ $\begin{pmatrix}3 & -5\\2 & -3\end{pmatrix}$ $\begin{pmatrix}-5 & -3\\7 & 4\end{pmatrix}$

12 $\frac{1}{3}\begin{pmatrix}8 & -9\\-5 & 6\end{pmatrix}$ $\frac{1}{2}\begin{pmatrix}6 & -4\\-4 & 3\end{pmatrix}$ $\frac{1}{4}\begin{pmatrix}2 & 3\\4 & 8\end{pmatrix}$ not possible

$\begin{pmatrix}-5 & 8\\2 & -3\end{pmatrix}$ $\frac{1}{14}\begin{pmatrix}6 & -2\\-5 & 4\end{pmatrix}$ $\frac{1}{6}\begin{pmatrix}-7 & -10\\-5 & -8\end{pmatrix}$

not possible.

13 $\begin{pmatrix}3 & -5\\-1 & 2\end{pmatrix}$ $\frac{1}{7}\begin{pmatrix}3 & -4\\-2 & 5\end{pmatrix}$ $\frac{1}{4}\begin{pmatrix}6 & -10\\-2 & 4\end{pmatrix}$

$\frac{1}{63}\begin{pmatrix}9 & -12\\-6 & 15\end{pmatrix}$ $\frac{1}{15}\begin{pmatrix}6 & -9\\-3 & 7\end{pmatrix}$ $\begin{pmatrix}0 & -1\\1 & -3\end{pmatrix}$

$\frac{1}{7}\begin{pmatrix}13 & -23\\-11 & 20\end{pmatrix}$.

14 (a) Both $\begin{pmatrix}6 & 8\\6 & 8\end{pmatrix}$ (b) $\begin{pmatrix}-2 & -6\\-6 & -2\end{pmatrix}$ $\begin{pmatrix}0 & 4\\-2 & 6\end{pmatrix}$

(c) Both $\begin{pmatrix}20 & 70\\18 & 72\end{pmatrix}$.

15 (a) Yes (b) No – only a particular case taken (c) Yes.

16 $\begin{pmatrix}3 & 9\\5 & 11\end{pmatrix}$ $\begin{pmatrix}1 & 1\\1 & 1\end{pmatrix}$ $\begin{pmatrix}12 & 33\\15 & 42\end{pmatrix}$ $\begin{pmatrix}14 & 29\\19 & 40\end{pmatrix}$ $\begin{pmatrix}19 & 40\\24 & 51\end{pmatrix}$

$\begin{pmatrix}9 & 24\\12 & 33\end{pmatrix}$.

(a) No (b) Yes (c) No (d) $(\mathbf{S} - \mathbf{T})\mathbf{T}$.

17 $\begin{pmatrix}7 & 8\\7 & 3\end{pmatrix}$ $\begin{pmatrix}37 & 42\\18 & 25\end{pmatrix}$ $\begin{pmatrix}13 & 4\\16 & 5\end{pmatrix}$ $\begin{pmatrix}40 & 11\\17 & 5\end{pmatrix}$ $\begin{pmatrix}15 & 23\\19 & 30\end{pmatrix}$.

18 All $\begin{pmatrix} 00 \\ 00 \end{pmatrix}$.

19 No Yes $\mathbf{AB} = 0$.

20 $\begin{pmatrix} 6 & 8 \\ 3 & 1 \end{pmatrix}$ $\begin{pmatrix} 4 & -3 \\ 2 & 5 \end{pmatrix}$ $\begin{pmatrix} 7 & 2 \\ 8 & 1 \end{pmatrix}$ $\begin{pmatrix} 8 & 2 \\ -3 & 5 \end{pmatrix}$ $\begin{pmatrix} 4 & -2 \\ -7 & 3 \end{pmatrix}$

21 (a) $\begin{pmatrix} 6 & 1 \\ 2 & 3 \end{pmatrix}$ $\begin{pmatrix} 4 & 7 \\ 1 & 2 \end{pmatrix}$ (b) $\begin{pmatrix} 38 & 10 \\ 25 & 7 \end{pmatrix}$ $\begin{pmatrix} 38 & 25 \\ 10 & 7 \end{pmatrix}$
(c) $\begin{pmatrix} 25 & 44 \\ 11 & 20 \end{pmatrix}$ (d) No (e) Yes.

22 (a) $\begin{pmatrix} 7 & 3 \\ 6 & 4 \end{pmatrix}$ (b) $\begin{pmatrix} -20 & 20 \\ 10 & -30 \end{pmatrix}$

23 (a) $\begin{pmatrix} 0 & 4\frac{1}{2} \\ 3 & 2 \end{pmatrix}$ (b) $\begin{pmatrix} -3 & -7 \\ -2 & -5 \end{pmatrix}$ (c) $\begin{pmatrix} 5 & -7 \\ -2 & 3 \end{pmatrix}$
(d) $\begin{pmatrix} -15 & 1 \\ 14 & -9 \end{pmatrix}$ (e) $\frac{1}{5}\begin{pmatrix} 9 & 3 \\ -6 & 7 \end{pmatrix}$
An infinite number eg $\begin{pmatrix} n & 0 \\ 0 & n \end{pmatrix}$

24 (a) $\begin{pmatrix} 1 & 1 \\ 1 & 1 \end{pmatrix}$ (b) $\begin{pmatrix} 1 & 1 \\ 2 & 2 \end{pmatrix}$ (c) $\begin{pmatrix} 1 & 0 \\ 0 & 0 \end{pmatrix}$ (d) $\begin{pmatrix} 0 & 0 \\ 0 & 2 \end{pmatrix}$
(e) $\begin{pmatrix} 3 & -7 \\ -2 & 5 \end{pmatrix}$ (f) $\begin{pmatrix} 3 & -7 \\ -2 & 5 \end{pmatrix}$

25 (a) $\begin{pmatrix} 1 & 1 \\ 1 & 1 \end{pmatrix}$ (b) $\begin{pmatrix} 2 & 2 \\ 1 & 1 \end{pmatrix}$ (c) $\begin{pmatrix} 1 & 0 \\ 0 & 0 \end{pmatrix}$ (d) $\begin{pmatrix} 2 & -1 \\ -7 & 4 \end{pmatrix}$
(e) $\begin{pmatrix} -4 & 2 \\ 14 & -8 \end{pmatrix}$

26 (a) $\begin{pmatrix} r^2 + s^2 & 2rs \\ 2rs & r^2 + s^2 \end{pmatrix}$
(b) $\begin{pmatrix} 2 & 1 \\ 1 & 2 \end{pmatrix}$ or $\begin{pmatrix} 1 & 2 \\ 2 & 1 \end{pmatrix}$ $\begin{pmatrix} 4 & 1 \\ 1 & 4 \end{pmatrix}$ or $\begin{pmatrix} 1 & 4 \\ 4 & 1 \end{pmatrix}$
$\begin{pmatrix} 4 & 2 \\ 2 & 4 \end{pmatrix}$ or $\begin{pmatrix} 2 & 4 \\ 4 & 2 \end{pmatrix}$ $\begin{pmatrix} 3 & 1 \\ 1 & 3 \end{pmatrix}$ or $\begin{pmatrix} 1 & 3 \\ 3 & 1 \end{pmatrix}$

27 (a) 6 −7 (b) 10 −11 (c) 4 5 (d) 2 −9
(e) 12 −7 (f) 8 3 (g) 8 −6 (h) 6 −6.

28 (a) (194 243 205) Total sales of each drink
(b) £17.32 (c) $\begin{pmatrix}130\\106\\161\\123\\122\end{pmatrix}$ Total daily sales

(d) (642) Total week's sales.

29 (a) $\begin{pmatrix}4 & 6 & 2 & 1\\4 & 4 & 3 & 3\\5 & 2 & 3 & 1\end{pmatrix}$ (b) $\begin{pmatrix}4 & 2 & 3 & 4\\4 & 4 & 2 & 2\\3 & 6 & 2 & 4\end{pmatrix}$

(c) £499 (d) £956.

30 (a) $\begin{pmatrix}5 & 5 & 5 & 5\\5 & 5 & 5 & 7\\5 & 5 & 5 & 5\end{pmatrix}$ (b) $\begin{pmatrix}3 & 4 & 3 & 1\\4 & 2 & 4 & 5\\3 & 2 & 1 & 3\end{pmatrix}$

(c) $\begin{pmatrix}3 & 2 & 3 & 5\\2 & 4 & 2 & 1\\3 & 4 & 5 & 3\end{pmatrix}$.

31 (a) £1 000 (b) £173 (c) £1 455.

32 (a) $\begin{pmatrix}50p\\46p\\36p\end{pmatrix}$ (b) $\begin{pmatrix}22p\\26p\\24p\end{pmatrix}$ (c) Total time for each wash.

33 (a) Total number of parts in each blend
(b) $\begin{pmatrix}244p\\312p\\412p\end{pmatrix}$ $\begin{pmatrix}41p\\39p\\41p\end{pmatrix}$

Exercise 18

1 (c) Translations 3 to the right and 4 up
4 to the left and 2 up

(d) $\begin{pmatrix}-3 & -3 & -3 & -3\\-4 & -4 & -4 & -4\end{pmatrix}$ $\begin{pmatrix}4 & 4 & 4 & 4\\-2 & -2 & -2 & -2\end{pmatrix}$

(e) $+\begin{pmatrix}5 & 5 & 5 & 5\\3 & 3 & 3 & 3\end{pmatrix}$.

2 (b) $T_1 + T_2 = T_2 + T_1$ (c) $\begin{pmatrix}5 & 5 & 5\\-2 & -2 & -2\end{pmatrix}$.

3 The final images are the same
Translations are commutative $\begin{pmatrix} 3 & 3 & 3 \\ 1 & 1 & 1 \end{pmatrix}$.

4 (c) x – axis y – axis the origin
(d) $M_1 + M_2 = M_3$.

5 M_1 and M_2 are anticlockwise M_3 is clockwise
Changes the sense of the labelling
Labelled as in the original
An odd number of reflections changes the sense of the labelling.

6 (b) Reflection in $y = x$ (c) Unchanged 1 2 3
Changed 4 5.

7 (c) Anticlockwise rotations about the origin of 90° 180° and 270°.

8 **R** gives an anticlockwise rotation of 90° about the origin
$\mathbf{R}^6$ is an anticlockwise rotation of 540°
Effectively $\mathbf{R}^6 = \mathbf{R}^2$.

9 (c) Enlargements, centre at the origin (d) $\begin{pmatrix} -1\frac{1}{3} & 0 \\ 0 & -1\frac{1}{3} \end{pmatrix}$

10 The lines joining corresponding points intersect at the centre of enlargement.

11 (c) Enlargements, centre at the origin
2:1 and 1:2 4:1 and 1:4
(d) Invariant 1 3 5 Variant 2 4.

12 (1) 3:1 2:1 1:2 (2) 9:1 4:1 1:4.

13 A stretch parallel to the x-axis
A stretch parallel to the y-axis.

14 Stretched parallel to the y-axis
Stretched parallel to the x-axis – becomes a ‘dachshund’.

15 Shear parallel to the x-axis – becomes a ‘pointer’.

16 (a) 2:1 (b) 1:1 (c) 4:1.

17 2 1 4 The value of the determinant gives the rate of increase in area.

18 10:1 3:1 3:1 16:1 10:1

19 The image is a straight line. Area = 0.
Value of determinant = 0 = change in area.

20 As 19.

21 Enlargement Stretch parallel to x-axis
Shear and stretch parallel to y-axis 9:1 3:1 3:1.

22 $\begin{pmatrix} \frac{1}{3} & 0 \\ 0 & \frac{1}{3} \end{pmatrix}$ $\begin{pmatrix} \frac{1}{3} & 0 \\ 0 & 1 \end{pmatrix}$ $\begin{pmatrix} 1 & 0 \\ -\frac{1}{3} & \frac{1}{3} \end{pmatrix}$.

23 (a) Reflection in y-axis
(b) Anticlockwise rotation of 90° about the origin
(c) Enlargement, centre at origin $\begin{pmatrix} 1 & 0 \\ 0 & -1 \end{pmatrix}$ $\begin{pmatrix} 0 & 1 \\ -1 & 0 \end{pmatrix}$
$\begin{pmatrix} -2 & 0 \\ 0 & -2 \end{pmatrix}$.

24 (a) Reflection in x-axis Anticlockwise rotation of 90° about the origin. Enlargement, centre at origin.
(b) Areas unchanged in **P** and **Q**
Directions in **R**
Shapes in **P Q** and **R**.

25 (a) **X Z** (b) **Y Z** (d) **W X**.

26 (a) $\begin{pmatrix} 1 & 0 \\ 0 & -1 \end{pmatrix}$ (b) $\begin{pmatrix} -1 & 0 \\ 0 & 1 \end{pmatrix}$ (c) $\begin{pmatrix} 0 & -1 \\ 1 & 0 \end{pmatrix}$
(d) $\begin{pmatrix} 2 & 0 \\ 0 & -2 \end{pmatrix}$ (e) $\begin{pmatrix} -3 & 0 \\ 0 & 3 \end{pmatrix}$.

Exercise 19

1 (a) $\begin{pmatrix} 0 & 1 & 1 \\ 0 & 0 & 1 \\ 1 & 0 & 0 \end{pmatrix}$ (b) Number of routes from the point
Number of routes to the point
(d) $\begin{pmatrix} 1 & 0 & 1 \\ 1 & 0 & 0 \\ 0 & 1 & 1 \end{pmatrix}$ 5 (e) 7 3.

2 (a) $\begin{pmatrix} 0 & 1 \\ 2 & 1 \end{pmatrix}$ 4 (b) $\begin{pmatrix} 2 & 1 \\ 2 & 3 \end{pmatrix}$ $\begin{pmatrix} 2 & 3 \\ 6 & 5 \end{pmatrix}$
(c) 8 (d) P – Q – P – Q in 2 ways
P – Q – Q – Q.

3 $\begin{pmatrix} 0 & 1 & 1 \\ 1 & 0 & 1 \\ 1 & 0 & 0 \end{pmatrix}$ $\begin{pmatrix} 2 & 0 & 1 \\ 1 & 1 & 1 \\ 0 & 1 & 1 \end{pmatrix}$ $\begin{pmatrix} 4 & 1 & 3 \\ 3 & 2 & 3 \\ 1 & 2 & 2 \end{pmatrix}$ 5 8 21 8.

4 5 9 16 Q — S — Q — S Q — P — Q — S.

5 5 7 13 (a) 7 (b) 7 (c) 1 Q and R R and P.

6 $\begin{pmatrix}1 & 1\\1 & 1\end{pmatrix}$ $\begin{pmatrix}2 & 2\\2 & 2\end{pmatrix}$ $\begin{pmatrix}4 & 4\\4 & 4\end{pmatrix}$ $\begin{pmatrix}8 & 8\\8 & 8\end{pmatrix}$ $\begin{pmatrix}32 & 32\\32 & 32\end{pmatrix}$ 128

7 $\begin{pmatrix}0 & 1 & 1\\0 & 0 & 1\\1 & 0 & 0\end{pmatrix}$ $\begin{pmatrix}1 & 0 & 1\\1 & 0 & 0\\0 & 1 & 1\end{pmatrix}$ $\begin{pmatrix}1 & 1 & 2\\1 & 0 & 1\\1 & 1 & 1\end{pmatrix}$ 5 7 **M** Yes.

8 $\begin{pmatrix}0 & 1 & 0 & 0\\1 & 0 & 1 & 0\\0 & 1 & 0 & 1\\0 & 1 & 1 & 0\end{pmatrix}$ $\begin{pmatrix}1 & 0 & 1 & 0\\0 & 2 & 0 & 1\\1 & 1 & 2 & 0\\1 & 1 & 1 & 1\end{pmatrix}$ Q to H

Q — M — C, deposit stone pick up oysters, —H deposit more stone and pick up lobsters — M, deliver oysters and lobsters.
Run MH in the morning and HM in the afternoon – or vice versa.

9 (a) $\begin{pmatrix}0 & 0 & 0 & 0\\1 & 0 & 0 & 0\\0 & 1 & 0 & 0\\0 & 0 & 1 & 0\end{pmatrix}$ (b) is 200 m lower than

(c) $\begin{pmatrix}0 & 0 & 0 & 0\\0 & 0 & 0 & 0\\1 & 0 & 0 & 0\\0 & 1 & 0 & 0\end{pmatrix}$

is 400 m higher than is 600 m higher than.

10 (a) $\begin{pmatrix}0 & 0 & 0 & 0 & 0\\1 & 0 & 0 & 0 & 0\\1 & 1 & 0 & 0 & 0\\1 & 1 & 1 & 0 & 0\\1 & 1 & 1 & 1 & 0\end{pmatrix}$ (b) $\begin{pmatrix}0 & 1 & 1 & 1 & 1\\0 & 0 & 1 & 1 & 1\\0 & 0 & 0 & 1 & 1\\0 & 0 & 0 & 0 & 1\\0 & 0 & 0 & 0 & 0\end{pmatrix}$

(c) $\begin{pmatrix}0 & 0 & 0 & 0 & 0\\0 & 0 & 0 & 0 & 0\\1 & 0 & 0 & 0 & 0\\2 & 1 & 0 & 0 & 0\\3 & 2 & 1 & 0 & 0\end{pmatrix}$ is above

The total number of ways in which the given book is separated by one or more books from another book.

11 (a) is $\frac{1}{3}$ as long as (b) is 9 times as long as.

12 (a) $\begin{pmatrix} 0 & 1 & 0 & 0 \\ 1 & 0 & 1 & 0 \\ 0 & 1 & 0 & 1 \\ 0 & 0 & 1 & 0 \end{pmatrix}$ (b) The relation is not reflexive

(c) $\begin{pmatrix} 0 & 1 & 0 & 0 \\ 1 & 0 & 1 & 0 \\ 0 & 1 & 0 & 1 \\ 0 & 0 & 1 & 0 \end{pmatrix} = \mathbf{B}$ The relation is symmetric.

13 (a) $\begin{pmatrix} 0 & 0 & 1 & 0 \\ 0 & 0 & 0 & 1 \\ 0 & 1 & 0 & 0 \\ 1 & 0 & 0 & 0 \end{pmatrix}$ (b) is on the left of

(c) $\begin{pmatrix} 0 & 1 & 0 & 0 \\ 1 & 0 & 0 & 0 \\ 0 & 0 & 0 & 1 \\ 0 & 0 & 1 & 0 \end{pmatrix}$

sits opposite (d) The relation 'sits opposite' is symmetric.

14 (a) $\begin{pmatrix} 0 & 1 & 0 & 0 & 0 & 0 \\ 0 & 0 & 0 & 1 & 1 & 0 \\ 0 & 0 & 0 & 0 & 0 & 0 \\ 0 & 0 & 0 & 0 & 0 & 0 \\ 0 & 0 & 0 & 0 & 0 & 0 \end{pmatrix}$ (b) is the child of

(c) is the father of the father of – this is *not* the same as grandfather of.

15 $\begin{pmatrix} 0 & 0 & 0 & 0 & 0 & 0 \\ 0 & 0 & 1 & 0 & 0 & 0 \\ 0 & 0 & 0 & 0 & 0 & 0 \\ 0 & 0 & 0 & 0 & 1 & 0 \\ 0 & 0 & 0 & 0 & 0 & 0 \\ 0 & 0 & 0 & 0 & 0 & 0 \end{pmatrix} \begin{pmatrix} 0 & 1 & 1 & 0 & 0 & 0 \\ 0 & 0 & 0 & 1 & 1 & 0 \\ 0 & 0 & 0 & 0 & 0 & 1 \\ 0 & 0 & 0 & 0 & 0 & 0 \\ 0 & 0 & 0 & 0 & 0 & 0 \\ 0 & 0 & 0 & 0 & 0 & 0 \end{pmatrix}$

$\begin{pmatrix} 0 & 0 & 0 & 0 & 0 & 0 \\ 0 & 0 & 0 & 0 & 0 & 1 \\ 0 & 0 & 0 & 0 & 0 & 0 \\ 0 & 0 & 0 & 0 & 0 & 0 \\ 0 & 0 & 0 & 0 & 0 & 0 \\ 0 & 0 & 0 & 0 & 0 & 0 \end{pmatrix}$

is the brother of a parent of i.e. is the uncle of.

16 (a) $\begin{pmatrix} 0 & 0 & 1 & 0 \\ 1 & 0 & 1 & 0 \\ 0 & 0 & 0 & 0 \\ 1 & 1 & 1 & 0 \end{pmatrix}$ (b) $\begin{pmatrix} 0 & 1 & 0 & 1 \\ 0 & 0 & 0 & 1 \\ 1 & 1 & 0 & 1 \\ 0 & 0 & 0 & 0 \end{pmatrix}$ is lower than

(c) $\begin{pmatrix} 0 & 0 & 0 & 0 \\ 0 & 0 & 1 & 0 \\ 0 & 0 & 0 & 0 \\ 1 & 0 & 2 & 0 \end{pmatrix}$

The number of ways in which the given mountain is higher than a mountain that is itself higher than the other given mountain.

17 (a) $\begin{pmatrix} 0 & 0 & 0 & 0 & 0 \\ 1 & 0 & 0 & 0 & 0 \\ 0 & 1 & 0 & 0 & 0 \\ 0 & 0 & 0 & 0 & 0 \\ 0 & 0 & 0 & 1 & 0 \end{pmatrix}$ (b) $\begin{pmatrix} 0 & 0 & 0 & 0 & 0 \\ 1 & 0 & 0 & 0 & 0 \\ 0 & 0 & 0 & 0 & 0 \\ 0 & 1 & 0 & 0 & 0 \\ 0 & 0 & 0 & 0 & 0 \end{pmatrix}$

(c) $\begin{pmatrix} 0 & 0 & 0 & 0 & 0 \\ 0 & 0 & 0 & 0 & 0 \\ 0 & 0 & 0 & 0 & 0 \\ 1 & 0 & 0 & 0 & 0 \\ 0 & 0 & 0 & 0 & 0 \end{pmatrix}$ $\begin{pmatrix} 0 & 0 & 0 & 0 & 0 \\ 0 & 0 & 0 & 0 & 0 \\ 1 & 0 & 0 & 0 & 0 \\ 0 & 0 & 0 & 0 & 0 \\ 0 & 1 & 0 & 0 & 0 \end{pmatrix}$

(d) $y = (x + 2)^2$ $y = x^2 + 2$.

18 (a) $\begin{pmatrix} 0 & 0 & 0 & 0 & 0 & 1 \\ 1 & 0 & 0 & 0 & 0 & 0 \\ 0 & 1 & 0 & 0 & 0 & 0 \\ 0 & 0 & 1 & 0 & 0 & 0 \\ 0 & 0 & 0 & 1 & 0 & 0 \\ 0 & 0 & 0 & 0 & 1 & 0 \end{pmatrix}$ (b) sits to the right of

(c) $\begin{pmatrix} 0 & 0 & 0 & 0 & 1 & 0 \\ 0 & 0 & 0 & 0 & 0 & 1 \\ 1 & 0 & 0 & 0 & 0 & 0 \\ 0 & 1 & 0 & 0 & 0 & 0 \\ 0 & 0 & 1 & 0 & 0 & 0 \\ 0 & 0 & 0 & 1 & 0 & 0 \end{pmatrix}$ $\begin{pmatrix} 0 & 0 & 0 & 1 & 0 & 0 \\ 0 & 0 & 0 & 0 & 1 & 0 \\ 0 & 0 & 0 & 0 & 0 & 1 \\ 1 & 0 & 0 & 0 & 0 & 0 \\ 0 & 1 & 0 & 0 & 0 & 0 \\ 0 & 0 & 1 & 0 & 0 & 0 \end{pmatrix}$

(d) $\mathbf{L}^3$ sits 2 places to the left of

(e) $\mathbf{L}^5$ = sits 5 places to the left of = sits on the right of.

Exercise 20

1 $\begin{pmatrix}3\\1\end{pmatrix}$ $\begin{pmatrix}2\\3\end{pmatrix}$ $\begin{pmatrix}2\\-2\end{pmatrix}$ $\begin{pmatrix}-5\\-1\end{pmatrix}$ $\begin{pmatrix}-4\\0\end{pmatrix}$.

2 $\begin{pmatrix}3\\1\end{pmatrix}$ $\begin{pmatrix}1\\-3\end{pmatrix}$ $\begin{pmatrix}2\\2\end{pmatrix}$ $\begin{pmatrix}-1\\-3\end{pmatrix}$ $\begin{pmatrix}0\\-4\end{pmatrix}$ $\begin{pmatrix}2\\0\end{pmatrix}$ $\begin{pmatrix}7\\-2\end{pmatrix}$.

3 $\begin{pmatrix}7\\6\end{pmatrix}$ $\begin{pmatrix}-1\\6\end{pmatrix}$ $\begin{pmatrix}9\\12\end{pmatrix}$ $\begin{pmatrix}17\\16\end{pmatrix}$ $\begin{pmatrix}8\\0\end{pmatrix}$ $\begin{pmatrix}11\\4\end{pmatrix}$.

4 $\begin{pmatrix}3\\5\end{pmatrix}$ $\begin{pmatrix}9\\-2\end{pmatrix}$ $\begin{pmatrix}12\\6\end{pmatrix}$ $\begin{pmatrix}10\\12\end{pmatrix}$ $\begin{pmatrix}-6\\7\end{pmatrix}$ $\begin{pmatrix}-2\\9\end{pmatrix}$.

5 $\begin{pmatrix}3\\2\end{pmatrix}$ $\begin{pmatrix}2\\-1\end{pmatrix}$ $\begin{pmatrix}-2\\3\end{pmatrix}$ $\begin{pmatrix}-3\\-2\end{pmatrix}$ $\begin{pmatrix}-6\\2\end{pmatrix}$ $\begin{pmatrix}-9\\-6\end{pmatrix}$.

6 7·2 at 56° 7·6 at 67° 8·9 at 27° 8·9 at 333°
7·8 at 130° 7·8 at 230° 7·3 at 344° 10·3 at 209°

7 5 53° 08′ 10 53° 08′ 10 306° 52′ 15 53° 08′
13 67° 23′ 13 292° 37′ 13 112° 37′
26 67° 23′ 17 28° 04′ 17 61° 56′.

8 (a) 5 (b) 13 (c) 10 (d) 29.

9 (a) $x = 5$ $y = -4$ (b) $p = 9$ $q = -2$.

10 (a) $p = 3$ $q = 3$ (b) $p = 6$ $q = 3$.

11 (a) 3 (b) ± 10 (c) ± 2.

12 (a) $\begin{pmatrix}-2\\-1\\-3\end{pmatrix}$ (b) $\begin{pmatrix}5\\-1\\7\end{pmatrix}$ (c) $\begin{pmatrix}-1\\3\\-1\end{pmatrix}$ (d) $\begin{pmatrix}1\\-3\\1\end{pmatrix}$

(e) $\begin{pmatrix}2\frac{1}{2}\\-\frac{1}{2}\\3\frac{1}{2}\end{pmatrix}$ (f) $\begin{pmatrix}-\frac{1}{2}\\1\frac{1}{2}\\-\frac{1}{2}\end{pmatrix}$.

13 (a) $\begin{pmatrix}7\\6\\9\end{pmatrix}$ (b) $\begin{pmatrix}5\\-1\\-2\end{pmatrix}$ (c) $\begin{pmatrix}8\\1\\1\end{pmatrix}$ (d) $\begin{pmatrix}10\\3\\4\end{pmatrix}$

(e) $\begin{pmatrix}-1\\5\\8\end{pmatrix}$ (f) $\begin{pmatrix}-3\\\frac{1}{2}\\1\end{pmatrix}$.

14 9 9 11 14 22 18.

15 (a) 31 m/s at 058° (b) 148 km/h at 333°
(c) 43 km/h at 202° (d) 79 m/s at 165°.

16 (a) 11·5 N 9·6 N (b) 18·8 N 6·8 N
(c) −9·0 N 15·6 N (d) 13·9 N −8 N
(e) −10·4 N −6 N.

17 (a) 95 N at 230° (b) 88 N at 052°
(c) 20·8 kgf at 195°.

18 (a) 49 N at 072° (b) 24 N at 066° (c) 37 N at 359°
(d) 45 N at 048°.

20 072° at 157 k.

21 (a) 051° at 128 k (b) 106° at 164 k (c) 255° at 101 k.

22 (a) 19·2 k at 141° (b) 24·6 k at 170° (c) 8 k at 298.

23 (a) 089° 132 k (b) 041° 147 k (c) 166° 108 k
(d) 255° 125 k.

24 36·5 km/h at 171° or 189°.

25 3·6 km/h at 56° to the bank downstream. 40 m.

26 (a) $\begin{pmatrix}6\\2\end{pmatrix}$ (b) $\begin{pmatrix}3\\6\end{pmatrix}$ (c) $\begin{pmatrix}-3\\-1\end{pmatrix}$ (d) $\begin{pmatrix}4\\3\end{pmatrix}$ (e) $\begin{pmatrix}2\\-1\end{pmatrix}$
(f) $\begin{pmatrix}2\\1\frac{1}{2}\end{pmatrix}$ (g) $\begin{pmatrix}1\frac{2}{3}\\1\frac{2}{3}\end{pmatrix}$.

27 (a) (10 12) (b) (−3 2) (c) (8 4) (d) (27 10)
(e) (−11 −2) (f) (4 2).

28 (a) $\begin{pmatrix}2\\3\end{pmatrix}$ (b) $\begin{pmatrix}3\\1\end{pmatrix}$ (c) $\begin{pmatrix}5\\4\end{pmatrix}$ (d) $\begin{pmatrix}2\frac{1}{2}\\2\end{pmatrix}$.

29 (a) $\begin{pmatrix}-2\\6\end{pmatrix}$ (b) $\begin{pmatrix}6\\-2\end{pmatrix}$ (c) $\begin{pmatrix}-6\\-2\end{pmatrix}$ (d) $\begin{pmatrix}-6\\-2\end{pmatrix}$.

30 (a) $\frac{1}{2}(\mathbf{p}+\mathbf{q})$ (b) $\frac{1}{4}(3\mathbf{p}+\mathbf{q})$ (c) $\frac{1}{4}(\mathbf{p}+3\mathbf{q})$
(d) $\frac{1}{5}(2\mathbf{p}+3\mathbf{q})$.

31 (a) $\begin{pmatrix}5\\5\end{pmatrix}$ (b) $\begin{pmatrix}6\\4\frac{2}{3}\end{pmatrix}$ (c) $\begin{pmatrix}4\\5\frac{1}{3}\end{pmatrix}$ (d) $\begin{pmatrix}4{\cdot}4\\5{\cdot}2\end{pmatrix}$.

32 (a) It is a parallelogram (b) it is a trapezium.

33 (a) $\begin{pmatrix}8\\2\\10\end{pmatrix}$ (b) $\begin{pmatrix}-3\\-2\\-4\end{pmatrix}$ (c) $\begin{pmatrix}7\\3\\9\end{pmatrix}$ (d) $\begin{pmatrix}1\\-1\\1\end{pmatrix}$ (e) $\begin{pmatrix}3\frac{1}{2}\\1\frac{1}{2}\\4\frac{1}{2}\end{pmatrix}$

34 (a) $\begin{pmatrix}2\\\frac{1}{2}\\2\frac{1}{2}\end{pmatrix}$ (b) $\begin{pmatrix}1\frac{1}{2}\\1\\2\end{pmatrix}$ (c) $\begin{pmatrix}3\frac{1}{2}\\1\frac{1}{2}\\4\frac{1}{2}\end{pmatrix}$.

Exercise 21

1 (a) AB = CD and AB//CD
(b) AB = BC and ABC is a straight line
(c) AB//CD and AB:CD = l:k (d) k = 0 and l = 0.

2 (a) $\frac{1}{3}\mathbf{p}$ (b) $\frac{1}{3}\mathbf{q}$ (c) $\frac{1}{3}(\mathbf{p} + \mathbf{q})$ (d) $(\mathbf{p} + \mathbf{q})$
(1) XY//AC and XY = $\frac{1}{3}$AC (2) XYCA is a trapezium.

3 (a) QR//PS and QR = $\frac{1}{2}$PS (b) PQRS is a trapezium
(c) $\mathbf{a} + \mathbf{b}$ $2\mathbf{b} - \mathbf{a}$ $\mathbf{b} - \mathbf{a}$ $\mathbf{b} - \mathbf{a}$
(1) QM//RS and QM = RS
(2) QRSM is a parallelogram.

4 (a) $\frac{1}{2}\mathbf{p}$ (b) $\frac{1}{2}\mathbf{q}$ (c) $\frac{1}{2}(\mathbf{p} + \mathbf{q})$ (d) $\mathbf{p} + \mathbf{q} + \mathbf{r}$
(e) $\frac{1}{2}(\mathbf{p} + \mathbf{q} + \mathbf{r})$ (f) $\frac{1}{2}(\mathbf{p} + \mathbf{q})$
(1) WX = ZY and WX//ZY
(2) WXYZ is a parallelogram.

5 (a) PQ//RS and PQ = $\frac{1}{3}$RS (b) a trapezium
(c) similar (d) 1:3.

6 $\mathbf{b}$ $2\mathbf{b}$ $\mathbf{a} + 2\mathbf{b}$ $\mathbf{b} - \mathbf{q}$ $\mathbf{a} + \mathbf{b}$
(a) BM//AN and BM:2AN (b) similar (c) 2:1.

7 $\mathbf{x} - \mathbf{y}$ $\mathbf{x} + \mathbf{y}$ $\mathbf{y} - \mathbf{x}$ $\mathbf{x} - \mathbf{y}$ $\mathbf{x} + \frac{1}{2}\mathbf{y}$
$-\frac{1}{2}\mathbf{x} - \mathbf{y}$ $\frac{1}{2}(\mathbf{x} - \mathbf{y})$ MN//SQ and MN – $\frac{1}{2}$SQ.

8 $\mathbf{q}$ $\mathbf{p} + \mathbf{q}$ $\frac{1}{3}(\mathbf{p} + \mathbf{q})$ $-\frac{1}{3}(\mathbf{p} + \mathbf{q})$ $\frac{1}{3}(2\mathbf{q} - \mathbf{p})$
$\frac{1}{3}(2\mathbf{q} - \mathbf{p})$
(a) VU = SW and VU//SW (b) a parallelogram.

9 $2\mathbf{p}$ $2\mathbf{p}$ $\mathbf{q} - \mathbf{p}$ $\mathbf{q} - \mathbf{p}$ AYCX is a parallelogram.

10 (a) $\mathbf{a} + \mathbf{b}$ $\mathbf{a} + \mathbf{b}$ SP = RQ and SP//RQ
(b) $\mathbf{a} - \mathbf{b}$ $\mathbf{a} - \mathbf{b}$ PQ = SR and PQ//SR
(c) a parallelogram.

11 (a) OPL is a straight line (b) OL = 20P
(c) $\mathbf{s} + \mathbf{q} = 0$ (d) a parallelogram (e) $\mathbf{p} + \mathbf{u} = \mathbf{t}$.

12 (a) $\mathbf{p} + \mathbf{q}$ (b) $-(\mathbf{r} + \mathbf{s} + \mathbf{t})$.

13 (a) VWX is a straight line VW = WX
(b) (1) VZ//XY and VZ = XY (2) a parallelogram
(c) (1) WX//ZY and WX = $\frac{1}{2}$ZY (2) a trapezium.

14 (a) (1) AB//DC and AB = $\frac{1}{2}$DC (2) a trapezium
(b) $\mathbf{p} + 2\mathbf{q}$ $2\mathbf{q} - \mathbf{p}$ $\mathbf{q} - \mathbf{p}$ $2(\mathbf{q} - \mathbf{p})$
(c) AE//BD and AE = $\frac{1}{2}$BD.

15 $\mathbf{y} \quad -\mathbf{y} \quad \mathbf{x} \quad -\mathbf{x} \quad \mathbf{x} + \mathbf{y} \quad -(\mathbf{x} + \mathbf{y}) \quad \mathbf{x} - \mathbf{y}$
$\mathbf{y} - \mathbf{x}$.

16 $\mathbf{x} + \mathbf{y} + \mathbf{z} \quad \mathbf{x} + \mathbf{y} - \mathbf{z} \quad \mathbf{x} - \mathbf{y} - \mathbf{z} \quad \mathbf{x} - \mathbf{y} + \mathbf{z}$.

17 $-\mathbf{p} \quad -\mathbf{q} \quad -\mathbf{r} \quad \mathbf{p} + \mathbf{q} \quad \mathbf{p} - \mathbf{q} \quad \mathbf{p} + \mathbf{q} + \mathbf{r}$.

18 (a) $\mathbf{p} \quad -\mathbf{q} \quad \mathbf{p} + \mathbf{q} \quad \mathbf{p} - \mathbf{q}$
(b) $-\mathbf{r} \quad \mathbf{q} - \mathbf{r} \quad \mathbf{p} + \mathbf{q} - \mathbf{r} \quad \mathbf{p} - \mathbf{r}$.

19 (a) $\mathbf{q} \quad -\mathbf{p} \quad \mathbf{p} + \mathbf{q} \quad \mathbf{r} - \mathbf{p} \quad \mathbf{r} - \mathbf{q} \quad \mathbf{r} - \mathbf{p} - \mathbf{q}$
(b) yes.

20 $\frac{1}{2}(\mathbf{r} - \mathbf{p} - \mathbf{q}) \quad \frac{1}{2}(\mathbf{p} + \mathbf{q} + \mathbf{r}) \quad \frac{1}{2}(\mathbf{q} + \mathbf{r} - \mathbf{p})$
$\frac{1}{2}(\mathbf{p} - \mathbf{q} + \mathbf{r})$.